FUNDAMENTALS OF
NATURAL GAS CONDITIONING

FUNDAMENTALS
OF
NATURAL GAS
CONDITIONING

R. N. Curry

665.73 C974

PennWell Books
PennWell Publishing Company
Tulsa, Oklahoma

Copyright © 1981 by
PennWell Publishing Company
1421 South Sheridan Road/P. O. Box 1260
Tulsa, Oklahoma 74101

Library of Congress cataloging in publication data

Curry, R. N.
  Fundamentals of natural gas conditioning.

  Bibliography
  Includes index.
  1. Gas, Natural—Purification. 2. Gas, Natural—Cleaning. 3. Gas, Natural—
Additives. I. Title.
TP754.C9   665.7'3   81–5077
ISBN 0–87814–162–6   AACR2

Printed in the United States of America

1 2 3 4 5 85 84 83 82 81

CONTENTS

ILLUSTRATIONS

TABLES

EQUATIONS

GRAPHS

ACKNOWLEDGMENTS

Much of the material in this book has been both condensed and borrowed from numerous publications and sources: equipment manufacturers, individual authors, handbooks, university seminars and short courses, government publications, and professional groups. All are aware that identical material found in any one published book, article, manual, or formal paper may well appear in several similar publications, and this work is no exception. Orginality in this kind of endeavor is only in the manner in which it is presented. Some, but not all of such publications, are the University of Oklahoma's *Gas Conditioning Conferences* and *Short Courses*, Professor L. S. Reid and W. E. Kinnebrew and the many gas industry people who contributed articles; *Handbook of Natural Gas Engineering*, Katz, Cornell, Kobayashi, Vary, Ellenbaas, Weinaug; *Monograph 8*, U. S. Bureau of Mines; *Gas Engineers' Handbook*, Industrial Press; *Gas Purification*, Kohl and Riesenfeld; *Hydrocarbon Processing; Oil & Gas Journal*; Welker Engineering Company; King Tool Company; Smith Industries; Perry Equipment Corporation; and Black, Sivalls, and Bryson, Inc.

INTRODUCTION

Gas conditioning is an all-inclusive definition. It is a term applied to numerous mechanical and chemical types of operational processes required to bring the gas up to pipeline quality. Among those processes and the ones of most concern to the gas industry are:

1. Gas cleaning
2. Additive injection
3. Desulfurization—sometimes referred to as gas purification, gas treating, etc.
4. Dehydration—also sometimes referred to as gas purification, gas treating, etc.

These processes are so interrelated that discussion of any one will necessarily attribute to the others, at least in part; thus the reason for some limited reference and discussion of each. While these processes can occur at any point along the pipeline system, the purpose and effort of this book is pointed toward the operation of a gas storage or production complex where these operational processes are required.

In a typical storage operation, these functions would initially occur in the sequence listed above. Gas cleaning will occur at several points before the product is delivered into the transmission system, i.e., at the wellhead, at the terminus of the gathering system, at the inlet and outlet to the dehydration and desulfurization system, and possibly again before compression.

PART I PRODUCT CHARACTERISTICS

Before discussing the different phases of gas conditioning, it seems appropriate that the physical makeup of the product—natural gas—be reviewed to establish a better understanding of what is being conditioned.

Ideally, and from the technical point of view, natural gas is made up of combustible components and exists as a family of hydrocarbons. Each member of the family contains a certain amount of carbon and a certain amount of hydrogen. That's why the name *hydrocarbons*. However, people who must process natural gas and operate the facilities through which it is conditioned realize, unfortunately, that the substance contains two other kinds of materials in addition to combustibles: diluents and contaminants.

Now, diluents are exactly as their name implies: they dilute. A diluent is a noncombustible gas that occurs in fairly small percentages in a typical natural gas. Because it is not combustible, it dilutes the total mixture and reduces its strength per unit volume, like adding water to bourbon. Among the most prevalent diluents are carbon dioxide, nitrogen, and water vapor. These compounds occupy space and are normally a part of the hydrocarbon environment. However, they have no BTU value.

There are many disadvantages in having diluents in the gas stream, most associated with horsepower, pipeline capacity, internal corrosion, and freezing. However, diluents may also produce desirable effects by reducing the energy in a cubic foot of gas while

1

keeping volume constant. This aspect is much like the cereal or filler in hamburger. In essence, the diluent content of typical natural gases poses no real problem because the quantities are normally small.

The problem constituents of natural gas withdrawn from production and storage fields are the contaminants. Gas conditioning just for their removal is the reason facilities are installed and operated. Contaminants and their source are almost infinite, but six are of most concern:

1. Water vapor above a certain concentration of about 5–7 pounds per million cubic feet (lb/MMcf).
2. All entrained free water or water in condensed form that may be present in the gas.
3. Any other fluid in liquid form that might be present in the gas, e.g., lube oil, scrubber oil, methanol, heavier-end hydrocarbons, or well inhibitors.
4. All solid matter that may be present (often defined as pipeline trash), including silica (sand), pipe scale, and dirt.
5. Acid gases, the main ones being hydrogen sulfide, even though it is a combustible, and, to a much lesser extent, carbon dioxide. In actual practice the only reason for removing CO_2 in concentrations of less than about 2–4% is because its removal occurs simultaneously in an MEA plant with the H_2S removal. Hydrogen sulfide in quantities of one or more grains per hundred cubic feet (1+ grains/Ccf) must be removed as required by the typical contract.
6. An amalgamation of any one of the other contaminants is normally referred to as sludge. But in pipeline parlance, it is called a number of even more descriptive terms.

At many installations, the unit of product/volume is the cubic foot, and it contains only the aforementioned combustibles and diluents less any contaminants. The control on diluent quantity is a contract condition. Product of less than a stated BTU per cubic foot value cannot be marketed. So what is a cubic foot of natural gas?

Normally by legal and contract definition, a cubic foot of gas is the volume that will occupy one cubic foot of space at an assumed atmospheric or barometric pressure of 14.7 and/or 14.4 psia (pounds per square inch absolute) and 60°F and is then further increased to a 14.73 pressure base. For gas conditioning purposes, a cubic foot of gas is assumed to be the volume that occupies one cubic foot of space at an atmospheric pressure of about 15 psia.

Typical pipeline-quality natural gas is composed of four main groups of compounds. Approximately 90–95% is made up of methane (CH_4). Another 2–4% is ethane (C_2H_6). Heavier ends through the C_6's constitute 2–3%. And 0.5–4% is CO_2 and N_2 as diluents with traces of water vapor and contaminants.

This, briefly, describes the physical conditions that apply to the general concept of a cubic foot of natural gas. Ideally, natural gas is one thing. But in a practical sense, it may be several things. So what are some of the characteristics of natural gas?

1. Technically and in static state (or if it were containable in an open glass jar), *natural gas—like the air we breathe—is imperceptible to the senses*. It has no odor, nor can it be seen, heard, or felt. But in actual practice, a person is usually aware of its presence. While we cannot smell the gas, we can smell the contaminants. And certainly we can usually see, feel, and hear gas expanding or escaping under pressure. Simply bear in mind that natural gas is like a phantom and, for reasons of safety, be aware that it has these noticeable characteristics.

2. *Natural gas is mass (it possesses weight) and occupies space.* It is sometimes difficult to form a tangible concept of something that cannot be seen nor felt; therefore, let's compare a volume of natural gas to something that can. The density of natural gas is equal to about 0.046 lb/cu ft. The density or weight of air is about 0.076 lb/cu ft. Thus, the specific gravity (sp gr) of natural gas as compared to air is 0.046/0.076, which equals 0.60. Therefore, a volume of natural gas weighs only 60% of an equal volume of air. Specific gravity, then, is the

ratio of the weight of gas compared to the weight of an equal volume of air.

Liquid fluids are not compressible in a practical sense. Certainly there is some compressibility. But if the pressure imposed on one cubic foot of water were increased and more water added up to 1,000 pounds, the volume would increase only a fractional amount. However, volumes of gaseous fluids are compressible almost proportionally to the increase in pressure. If a 1-cu ft vessel contains one standard cubic foot of gas at 15 psia, then at 30 psia it would contain 2 standard cubic feet or one mile of 24-in. pipe at 904 psia would contain 1.2 million standard cubic feet of gas. It would weigh about 55,000 lbs, or 28 tons more than if it were empty and at zero psi.

The equation for this process is:

$$SCF = (\pi r^2/144)\left(\frac{psig + 14.4}{14.73}\right)(L_f)\left(\frac{60 + 460}{°F + 460}\right)y$$
$$= 1,200,000$$

where:

$$\pi = 3.1416$$
$$r = \text{½ pipe ID (internal diameter)}$$
$$L_f = \text{length of pipe, ft}$$
$$°F = \text{gas temperature}$$
$$y = \text{supercompressibility}$$
$$\text{total wt., lb} = SCF \text{ (wt./cu ft)}$$
$$= 1,200,000 \text{ (0.046)}$$
$$= 55,200$$

3. *Energy*. This term has become more prominent in gas industry language in recent years and will become moreso since many gas companies have revised their tariffs (contracts) wherein they buy, sell, or exchange units of energy (BTUs) in place of cubic feet. The mechanics for measurement re-

main essentially the same, but the billing unit is adjusted for energy transferred instead of cubic feet. BTUs (British thermal units), or energy required, are the basis for design of much of the mechanical equipment used in a gas conditioning process, i.e., reboilers, heat exchangers, still columns, and fan units. One BTU is the amount of energy required to increase the temperature of one pound of water one degree Fahrenheit. The energy in one cubic foot of a typical natural gas is about 1,000 BTUs, or one pound of gas contains about 22 cu ft or 22,000 BTUs.

4. Natural gas is *one of the more stable flammable gases*. It is flammable within the limits of a 5–15% mixture with air and its ignition temperatures range from about 1,100–1,300°F. This is as compared to the less stable gases. For example, H_2S is flammable within the limits of about 4–46% in air and acetylene within about 2–80%, both with equally lower ignition temperatures.

5. As noted previously, natural gas, because it is typically 80% or more methane (CH_4), *is lighter than air*, whereas bottle gas (propane and butane) is heavier than air. Therefore, natural gas disperses rapidly in the atmosphere, and the heavier gases hover close to the ground. Consequently, heavy gases are less safe to use in places where a leak may occur. Hydrogen sulfide is also heavier than air. However, the less dense a gas, the more difficult it is to contain it or to prevent leakage.

6. *Leakage and gas blown to the atmosphere* around a storage field or a gas conditioning facility can be significant and quite costly. In instances such as blowing well syphons or drips, it is simply a part of the operating cost. In other cases, it is leakage from a valve that has been blown too many times. For reasons of evaluating such gas usage or leakage, knowing just how much product is escaping is helpful. The most accurate means to do this is to meter it, but in many instances this is neither practical nor convenient. There are, however, certain methods for arriving at the questionable volumes

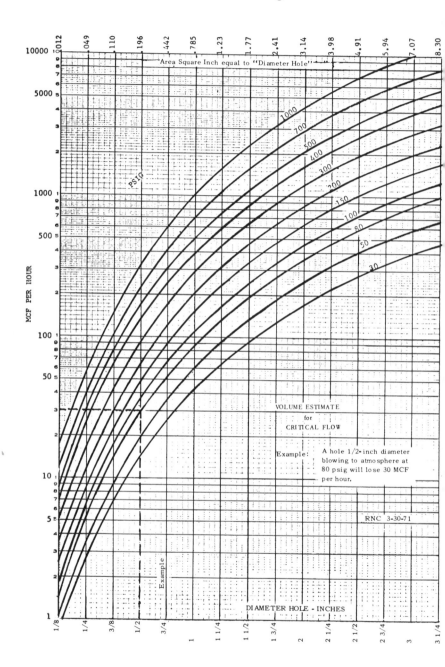

1–1 Volume estimate for critical flow

Valve leakage volumes can be reasonably approximated with a pitot tube arrangement and computated as shown:

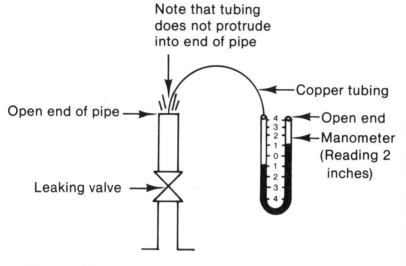

Water in U-tube
$$Q = 34.69 \; D^2\sqrt{W}$$

Mercury in U-tube
$$Q = 128 \; D^2\sqrt{M}$$

where:
  Q = Mcf / day
  D = diameter of pipe
  W = in. of water;
  M = in. of mercury

Example: A 2-in. blowoff is leaking and the impact pressure reads 9 in. of water. The leakage rate is calculated as:

$$Q = 34.69 \; (4) \; \sqrt{9}$$
$$Q = (34.69) \; (4) \; (3)$$
$$Q = 416 \; \text{Mcf / day}$$

1–2   Flowing gas volume determination with a pitot tube with minimum effort. Where the area of the opening can be determined, such as the opening in a plug valve or the area of a hole in a line, the volume escaping can be determined graphically, as shown in Fig. 1–1.

A point to remember is that most plug valves have an area opening equal to about 60–65% the size of the valve, or a 1-in. valve has a flow-through area of approximately 0.45 square inches (sq. in.),

where:

$$\text{Area, sq in.} = \pi r^2$$
$$= 3.14 \, (0.5^2) \, (0.60)$$
$$= 0.45$$

A ball valve or gate valve is usually considered a full-opening or through-conduit valve, and its area for a 1-in. valve is approximately ¾ sq in. Note that Fig. 1–1 includes two reference scales, one for diameter where the opening is presumably circumferential and the other for area to determine the volume where the opening is square or rectangular.

To determine the volume of gas escaping from a leaking valve in a closed position or a similar circumstance where no actual area of opening can be determined, a relatively simple pitot tube can be made from miscellaneous items usually available around a work location. This requires a 2-in. nipple about 12 in. long, a short length of ¼-in. tubing, and a glass-tube manometer to measure the impact pressure and arithmetically convert it to a volume. (Fig. 1–2).

This discussion has pointed out some of the characteristics of natural gas for safety and handling purposes. By no means is it all-inclusive. Rather, it has been intended merely to describe a condition for personal awareness.

PART II GAS CLEANING

Gas cleaning techniques and the design of vessels for cleaning are as varied as the number of companies that manufacture them. Some phase of the gas cleaning process usually occurs initially at the wellhead in the form of drips, filters, and syphons. This may happen again along the gathering system before the gas arrives at the plant location. At the plant, the gas is further cleaned before it contacts the solutions: TEG and MEA. There are two reasons for all of this cleaning: First, we need to be sure the gas can flow from the wellhead to the processing area in the most efficient and practical manner without freezing or reducing gathering-line efficiencies. Second, the gas must be cleaned before contacting the purifcation solutions to help prevent contamination of the solutions and consequently contaminating and fouling the regenerative system in general. Thus, gas cleaning and solution cleaning are equally important.

Gas cleaning is the oldest of the gas conditioning processes and has developed by stages only when improved performance was absolutely necessary. Each advance has resulted in a further reduction in size and quantity of matter that can be removed from the gas stream.

In broad terms, pipeline trash is anything in the flowing fluid other than gas and is an amalgamation containing parts of both liquids and solids, sometimes referred to as sludge or gook. The liquids are usually water, some heavier-end hydrocarbons, methanol, lube oil, glycol, and amine. Liquids collect in low spots or in sags in the gathering system and are moved downstream in slug flow by the gas while additional liquid collects. Solids can be either simple or complex mixtures, consisting of—among other things—drilling mud, construction dirt, mill scale, welding slag, sand (silica), steel cuttings, plug valve grease, and hydrates.

The liquids and solid trash are reasonably easy to remove with drips, syphons, and scrubbers. (Note: Some authorities distinguish between scrubbers and separators, contending that separators do not use auxiliary media—oil, for example—while scrubbers do. Other than for technical description, scrubbers and filters can generally mean the same thing for most operating purposes.) It is, however, important to remove these contaminants as they accumulate, either by manual or automatic means. Obviously, if the vessel fills and another slug comes through, it will continue in the flowing stream. Thus, the larger-sized solids and the greater quantities of slugs are problems. But they can be contained and removed.

The contaminants that present the greater challenge are the minute solids and liquid particles suspended in and being carried by the velocity of the flowing gas stream. These contaminants are called *particulate matter*, and the liquids are sometimes classed as *aerosols*. The quantity of particulate matter in a gas stream is usually expressed as pounds per million cubic feet. It has been said that a clean gas transmission system might average 2 lb/MMcf, but under certain conditions it may be as high as 300-400 lb/MMcf. In a gathering system, the quantity is assumed to be somewhere between these two figures at any given time, depending upon reservoir level, gas velocity in the well and tubing, and past experience. It is during these periods that extra attention must be given to removing as many of these submicronic particulates as possible.

In order to establish a concept of particulate size, the micron scale is used (25,400 microns equal 1 in.). The human eye cannot normally see a particle of less than 10 microns in diameter without magnification. An average strand of human hair is about 100 microns in diameter. A 400-mesh screen will enable particles of 37 microns to pass through it. Cigarette smoke is a dense dispersion of particulates with diameters ranging from ¼–1 micron. Fig. 2–1 is a scale illustrating the micron dimension concept.

Oil-bath and dry-type properly designed and maintained scrubbers can remove particulate matter down to 4 microns. Cartridge-type and sock-filter coalescer type cleaners can remove particulate matter down to 1 micron in diameter. However, the capability of removal brings about a proportional increase in maintenance. Obviously, if the cleaning element—whether it be adsorptive or absorptive—has accumulated its capacity, there are only three

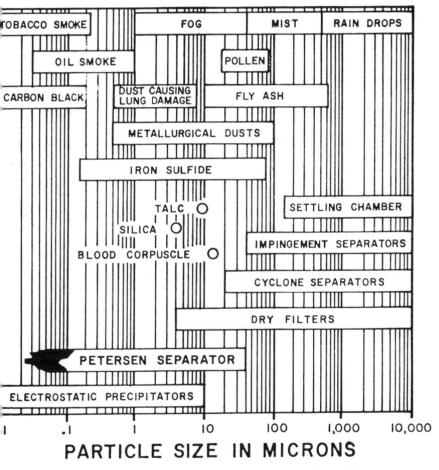

2–1 Typical particles and separation equipment *(after Holm)*

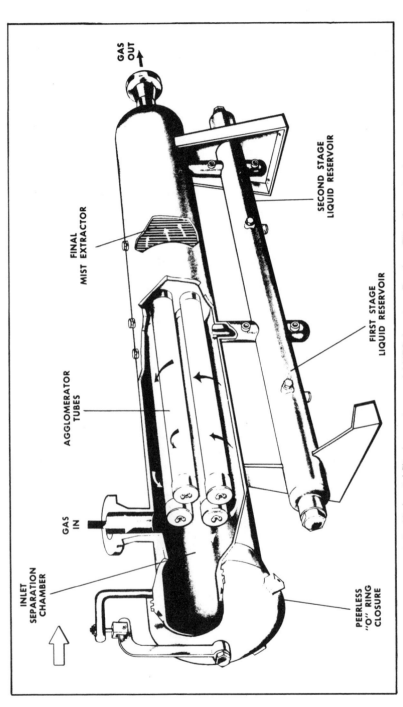

2–2 Horizontal filter/separator, model HFS *(courtesy Peerless Filter Separators)*

possible results, any one of which is detrimental to the cleaning process. First, if it has accumulated its capacity, then all excess will somehow enter (bypass, sloughing, etc.) the downstream system which we're trying to keep clean. Second, the cleaning element will create such a pressure differential that it may rupture or become dislodged, thereby adding its own matter to the stream. Last, the element can become so impregnated with particulate matter that it will decrease or totally interrupt the flow or process. Certainly, all three of these conditions have been encountered, but the first is probably the most prevalent and the most ignored.

There are various methods, none of which have been too successful, for measuring the quantity of particulate matter flowing in a gas or liquid stream. Millipore discs are the most common means of measurement. However, this aspect is of small importance to the gas conditioning processes normally used by the industry. It could be important for design and specification purposes, but in current day-to-day total field operations as much matter as is conceivably possible must be removed from the gas and process solutions with the physical equipment installed for that purpose. Gas cleaning equipment that is not properly maintained and serviced may induce results directly opposed to the system's purpose, particularly if it releases slugs of contaminant in a short period of time. The system could probably cope better with a constant but much lesser contaminant quantity on a consistent basis than an immediate impact of this material.

There are about five basic types of gas cleaners the industry has used at various places. The overall gas conditioning process in one fashion or another utilizes all types: gas-liquid separators, oil-bath scrubbers, dry scrubbers, cartridge-type scrubbers, and dust filters. In some instances the single vessel may utilize one or more of these methods.

For purpose of clarification, certain often-used terms are defined as follows:

Adsorption occurs when matter accumulates on a surface such as a baffle plate.

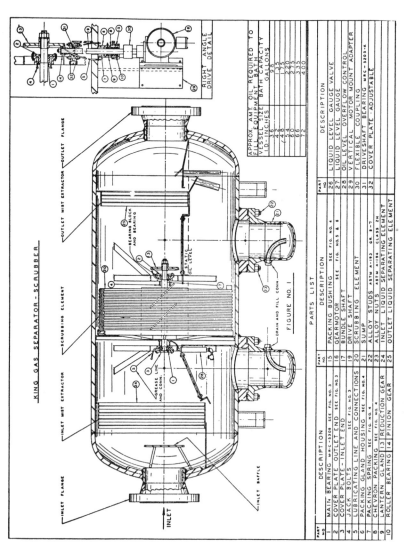

2-3 King gas separator-scrubber *(courtesy King Tool)*

Absorption is a blotting action—a cartridge filter.

A *coalescer-type unit* can be thought of as a piece of corrugated metal where the gas would strike the surface and the entrained liquids would form together in the recesses and gravitate to a liquid collection zone.

Impingement is the process of gas striking a substance such as steel wool, separating, and changing direction of flow numerous times. In doing so, the fine particles of liquid accumulate and gravitate downward to the liquid collection reservoir.

The most common unit is the gas-liquid separator, usually cylindrical in design, both horizontal and vertical. It is also known as a line scrubber, inlet scrubber, interstage scrubber, and outlet scrubber. Scrubbers of this type are designed with three basic functions:

1. *An inlet area* where gas is separated from the contaminants by means of gravity settling, centrifugation, and impingement on wet surfaces, and the gas moves to the mist extracting section.
2. *A liquid collecting section* at the lower end of the vessel to utilize gravity flow. It is equipped with a means of removing the liquid and sludge either manually or automatically.
3. *A mist-extracting section* located in the upper portion of the vessel that houses one or more mist extractor elements that impinge, coalesce, and recover additional particulate matter that escaped the inlet area. The mist extractor section must include a way of conducting the additionally recovered particulate matter to the liquid collection section through a conduit or some other means that will prevent further chance for reentrainment.

Scrubbers of this type are the most prevalent in the gas industry. They are advantageous because of simplicity of design and minimum operational and maintenance requirements.

The term "mist extractor" has appeared in the previous discussion and probably warrants further explanation. This item and its

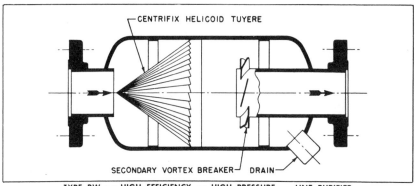

TYPE RW — HIGH EFFICIENCY — HIGH PRESSURE — LINE PURIFIER

2–4 Centrifix type RW high-pressure line purifier *(courtesy Centrifix, division of Burgess Industries)*

function are vital parts of most gas and solution cleaning processes. Where clean gas is the objective, it removes fine liquid particulate matter; where solutions are concerned, it is a conservation technique. There are variations of mist extractors, some of which employ the coalescer principle accompanied by high velocity and resulting impingement. However, in most gas-cleaning processes a mist extractor is really nothing more than a steel sponge—hopefully, a stainless steel sponge—or more generically steel wool. The material is usually knitted into a symmetrical mesh fabric and formed into a 4–6-in. thick pad.

A mist extractor is not to be confused with a filter. Its only function is to impinge and coalesce liquid particulates. The air or void space in a mist extractor is about 90–99% and utilizes very small diameter wire in the range of 0.003–0.011 in. Maximum efficiency is obtained at a velocity flow of 5–10 feet per second (fps). This velocity is needed to force the gas to impinge on the wires and, as a result, to coalesce into droplets that gravitate from the pad into the liquid accumulator. Greater velocities flood the pad and consequently reentrain the liquid. Lesser velocities permit the liquid to circumvent the pad and avoid impingement. Mist extractors of this type can remove liquid particulates in the range of 1–3 microns. These units cannot handle high concentrations of solids because of the pressure differentials that may occur. For the same reason, pressuring a vessel equipped with a mist extractor must be accomplished at a rate that will not damage the device. Recognizing there are other designs of mist extractors such as the fiberglass cartridge, the industry's most common application in gas storage and production fields is the mesh pad. In some instances a device called an Aerotec tube is used; it will remove both solid and liquid particulates as small as 4 microns in diameter. This device simply increases the velocity by flowing through a series of small tubes and radically changing direction of flow.

The next type of scrubber is the oil-bath type, used extensively at all types of installations where the gas must be cleaned. There are many different designs of this principle, but the common objective is to cause dispersed solid particulates to impinge on and cling to an oil-wet surface. The surface is continuously flushed with oil, thereby washing the dirt down into a sump where it is supposed to settle by gravity, thus providing relatively clean oil for recirculation over the contact surfaces. The scrubbed gas then passes through a mist extractor and back into the system.

There are two precautions for operating this type unit. First, the oil must be clean. Second, the proper oil level must be maintained. Excessive level will induce carryover, in turn overloading the mist extractor and the result may well be that the gas is leaving the scrubber with more contaminant than when it entered.

Dry scrubbers impart high velocities with resultant centrifugal force to effect the separation of dirt and liquids from the gas. They are reasonably effective down to particulate matter of 4 microns. The capacity requirement is such that differential pressure around 3 psi or slightly above must be maintained. As the velocity decreases, the effectiveness of the separation process is diminished.

Probably the most effective cleaner in gas-industry use is the cartridge type, first introduced about 20 years ago. This cleaner employs a parallel stacking arrangement of the cartridge and, depending upon the manufacturer, takes on many different design configurations, most of which in the larger sizes are horizontally installed. Where liquids are to be separated, coalescers and mist

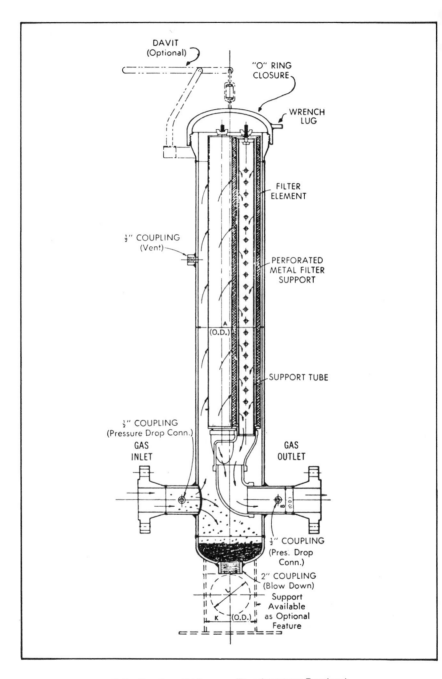

2–5 Peerless N-Line gas filter *(courtesy Peerless)*

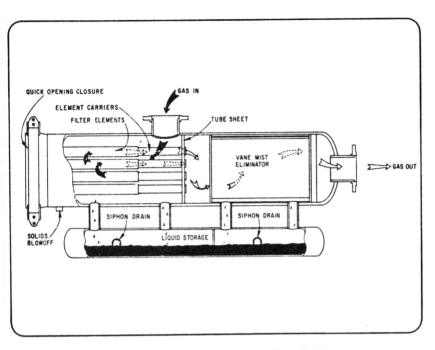

2–6   Peco filter separator unit *(courtesy Peco Robinson)*

extractors are a part of the unit. Cartridge-type cleaners can remove particulate solid matter down to 0.3 microns, a tremendous improvement over efficiencies for oil-bath and dry scrubbers. In gas conditioning solution field application, these units require continuous maintenance, back washing, and cartridge replacement.

Do not get the idea that cartridge-type filters are a panacea. In fact, they may cause additional problems, particularly in areas where the contaminants produce a coating effect. Here, depth penetration of the particulate matter is no greater than the sock container. Such contaminants may be certain types of corrosion inhibitors, paraffins, and varnishes. This question is discussed further under Part V, Desulfurization.

One other type of scrubber deserves recognition: the electric precipitator. At this time its application to large-volume, high-pressure gas cleaning has been limited and only a few have been installed. Because of the environmentalists, we hear much of this

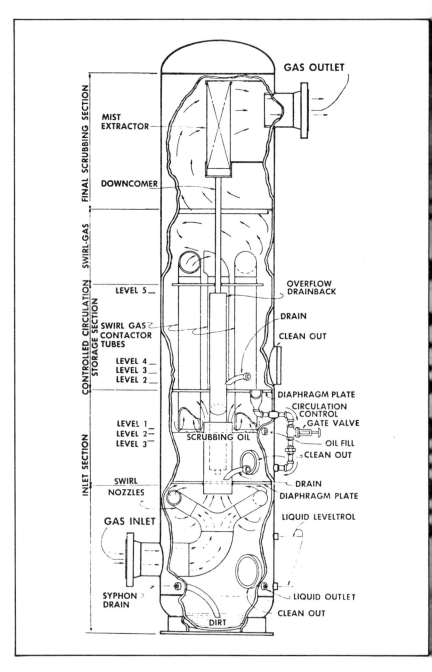

2–7 Peerless swirl gas cleaner *(courtesy Peerless)*

type of cleaning today. It may possibly hold some future for on-stream cleaning of natural gas. Very basically, it induces an electrical charge that attracts the particulate matter, from whence it is further accumulated in a depository.

This information on gas cleaning has emphasized the necessity for this operation at storage and production locations. If cleaning is important on main transmission systems, it is many times more important for storage and production gas. The contamination and fouling of purification plant solutions and equipment will occur, decrease, and increase just about proportional to the quality of the product which it purifies.

PART III ADDITIVE INJECTION

Gas withdrawn from a storage or production well frequently needs the addition of inhibitors to prevent corrosive or chemical action in the gathering system. Also, some storage reservoirs require that inhibitors be added to the gas being injected into the reservoir for downhole protection to the casing, tubing, and sand face.

From a day-to-day operational point of view, though, priority must be given to preventing ice and hydrate formations in the gathering system during the high-demand periods to ensure an uninterrupted flow of gas from the wellhead to the terminus, where the total flow may then even be further conditioned by cleaning, dehydration, and sometimes desulfurization. There are several different agents—methanol, the glycols, or ammonia derivatives—that could be used to prevent this from happening, but the most prevalent and successful agent is methanol. It disperses well in the gas stream, is readily available in bulk quantity, does not require recovery, and is usually the least expensive.

Water content of natural gas will be reviewed later in conjunction with dehydration, but at this point we may safely presume that gas withdrawn from a geological formation is saturated with water vapor at operating temperatures and pressures. We may also assume that it carries a substantial quantity of entrained free water, or water in condensed form. (This is not 100% true, particularly during the first part of a withdrawal cycle from a storage reservoir where the injected gas was dry.) Condensed water will freeze at 32°F and hydrates will form at pipeline flowing gas temperatures up to 60 or 70°F, depending upon the pressure, turbulence, etc. Fortunately, methanol will prevent either from happening.

The quantity of methanol required to prevent freezing and hydrate formation may be derived from the equations and data in the *Gas Engineers' Handbook* (Industrial Press), Section 4, p. 4/74 of the 1977 edition, or Chapter 8, p. 4/172 of the 1965 edition. However, the actual quantity required may not be as great as indicated.

One of the larger gas transmission companies operates numerous storage and production fields composed of several hundred wells and about the same number of miles of gathering system. Flowing gas temperatures and pressures during the winter months and the withdrawal season vary from 35–60 °F. and 100–1,000 psig. During the storage injection period, the pressures are about the same, but temperatures are considerably higher. Three of the storage reservoirs are aquifers, and the gas withdrawn from them is sour, containing H_2S up to 100 grains/100 Ccf. During the winter months, the gas at the wellhead must be conditioned with methanol; during the storage injection season, downhole inhibitors must be added.

Several years ago, companies were experiencing unsatisfactory results in both areas: freeze-offs were occurring during the winter, and the inhibitor was not dispersing properly downhole during the storage cycle. Both problems were directly attributable to the gravity-type injection system being used. It consisted of a capacity tank fabricated from pipe with welding caps, a sight glass, and valve manifolding to maintain equal pressure across the chemical and to regulate its entry into the flowing gas stream. Injection for the most part was sporadic, either dumping the additive in a short time or failing to feed at all. The solution was obvious. The methanol and inhibitor had to be added to the gas at consistently uniform rates. The problem was corrected by replacing all of the fabricated gravity-type units with a pneumatically powered chemical injection pump.

The next phase was to determine just exactly how much methanol was required at each wellhead to prevent blockage. By using the equations noted earlier in this text, it was derived that 2–3 gal/MMcf would be needed, depending upon the particular location. The program was initiated by injecting the maximum amount of methanol indicated as required. But in the interest of economy, that amount was gradually reduced to a point where freeze-offs began to occur and then slightly increased. In the final analysis, it was found that approximately 1 gal of methanol per MMcf injected at the wellhead at a constant rate (the aquifers and greater H_2O content fields requiring about 1.25 gallons) would prevent freezing and hydrate formations in the gathering systems. Proof of this conclusion was continuous, uninterrupted operation without ice or hydrate blockage since that time. Inhibitor addition during the summer months was equally effective.

There were other benefits gained from this program, mainly economical. Field personnel could operate on a regularly scheduled basis rather than making random maintenance trips to injection sites. Exact methanol requirements could be anticipated prior to the withdrawal season and inventories could be replenished. Methanol, like other products, is less expensive per unit volume if purchased in large quantity, and this reflects favorably upon operating budgets. One of the less intrinsic benefits is the at least partial relief from anxiety that every operator experiences during the colder months.

This method should not be construed as a recommendation that, when using the chemical injector pump, one gallon of methanol/ MMcf is the correct amount to prevent stoppage. Rather, it is a suggestion that each company should conduct a similar type of evaluation to determine its specific requirements. One gallon/ MMcf might be a good place to start.

Obviously, individual wells do not produce equal volumes of gas. Consequently, some wells require far greater quantities of methanol than others. Where measurement occurs at the wellhead, as is the usual case in a production field, there is no problem in correlating methanol to MMcf production. But storage wells are not always individually metered, and their daily production will vary with reservoir levels, rate of total demand on the reservoir, permeability of the sand, and probably many other considerations. If a withdrawal volume cannot in some manner be established for a particular well by installing a temporary meter, the information may be obtained from open flow and back pressure deliverability tests. The technical staff in the production or storage groups can provide that information.

There are a variety of manufacturers marketing chemical injection pumps for applications of this type, and any one may be equally as satisfactory as another. By comparison with other types of equipment normally required for a production operation, the cost of an injector pump might be considered insignificant, and an in-service evaluation be made of several different kinds. In selecting a particular pump, consider availability of the unit and repair parts, cost, observed performance, power gas consumed, and frequency of repair required. Any pump utilizing brass or copper trim should be avoided, particularly if it is to be used in a sour-gas application. In fact, any component, fitting, tubing, etc., containing brass or copper should be avoided if its application is in a mercury or H_2S environment—even moreso if it is a pressure-containing device. Either compound, Hg or H_2S, will cause a physical breakdown of the metal, and the result is much like stress corrosion. Regardless of final selection, the pump will work quite satisfactorily at certain locations, while at others there will be problems. This is inevitable, like the Ford-Chevrolet syndrome: assign a Ford vehicle to an employee who prefers a Chevrolet and the result is disastrous—and vice-versa.

All of these injector pumps are positive-displacement type, meaning that they will pump the quantity of inhibitor per stroke whether it is discharging to a 10- or 1,000-psig system. The rate at which any pump works is regulated by mechanical adjustment. The rate can actually be measured by pumping to atmosphere for a short period and measuring with a vessel. Fig. 3–1 is a reasonable approximation for converting drops/minute to gallons/day.

The manifolding and installation arrangement for these pumps is quite critical for two very important reasons. First, the H_2O-saturated operating media wellhead power gas cannot be allowed to freeze because of the pressure reduction required to operate the pump. Second, one needs to be certain that the methanol is in fact

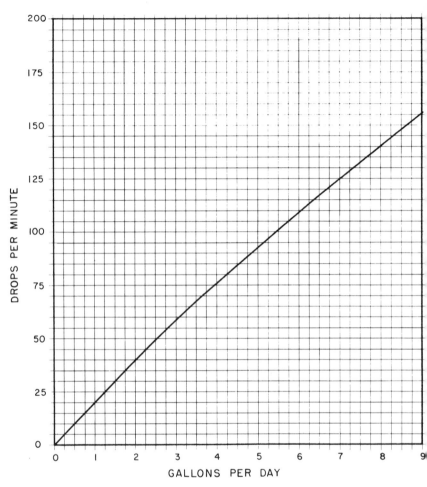

3–1 Conversion chart from gallons per day to drops per minute

entering the main gas stream and is not being totally entrained in the pump's power supply. Ideally, the power gas will contain just enough absorbed methanol to prevent freezing, while allowing the far greater quantity to enter the mainstream. Trial and error concluded that the total methanol supply must be routed downward into the running end of a vertically installed tee, and the pump's power or operating gas then taken from the bull-headed side of the tee. The trick is simply that the tee functions as a sort of mixing valve. With injection rates of 1–5 gallons per day, any tee larger than 1 in. doesn't provide sufficient methanol in the power gas to prevent freezing, and smaller tees absorb too much methanol in the power gas. Certainly, where great quantities of methanol must be injected into a well producing large volumes of gas, system components must be proportionately increased in size: storage tank, manifolding, pump, and the mixing valve tee. This will require some experimentation.

Figure 3–2 is a schematic drawing of recommended manifolding and overall installation for a typical methanol injection system using one type of pump. It also works well with pumps by different manufacturers. One item of most importance is the rupture disc (18), located in the pump's discharge piping. As noted earlier, these are positive-displacement pumps (sometimes called plunger pumps), and they sense no buildup of discharge pressure. Consequently, if pumping into an inadvertently closed valve or system, the result could be catastrophic without the rupture disc. The discs in the figure are ¼-in. and may be obtained from any supplier. They are like the type used on compressor engine high-pressure lube systems.

The manifolding in the figure is for methanol injection only. Where such installed pumps are also used for other kinds of inhibitor injection, the pump's power must be rearranged whereby that supply is obtained upstream of the inhibitor injection point. Otherwise, a part of the particular chemical will be absorbed in the power gas. While this is the intent for methanol, other compounds may tend to cause an operational malfunction of the pump and foul the regulator. Simply remember that when adding methanol, the power supply is from the point of injection; but when adding other inhibitors, the supply is from a point upstream of injection.

At locations with some kind of metering device, additive injection can be achieved at a proportional-to-flow rate, thereby eliminating any manual adjustment whatever. In the past, methanol has been a relatively inexpensive product and did not seemingly warrant a more sophisticated and complex injection system. However, where a methanol injection system is being initially contemplated,

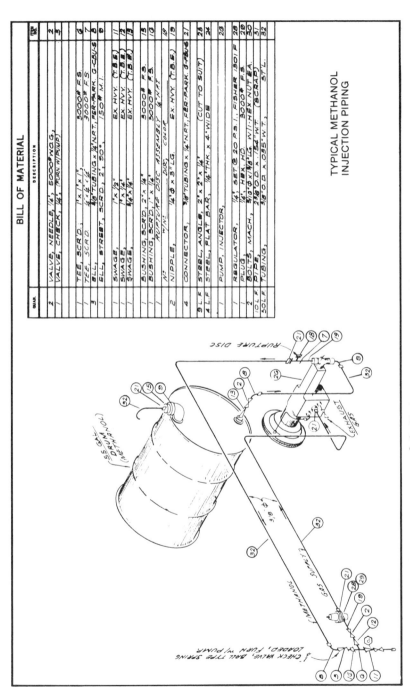

3-2 Typical methanol injection piping

at least explore the possibility of a proportional-to-flow system in light of current and projected operating and methanol costs. This will require additional instrumentation.

Without a doubt, the pump systems are more effective and efficient than the gravity-type systems, but many of the latter are still used. And where continuity of flow is not critical or where there are so few installations required that operational functions are minimal, the gravity system can be satisfactorily utilized. They do have one advantage: the complete absence of any trigger work or motion machinery. Fig. 3–3 is a sketch of one such device capable of holding about 5 gallons of liquid. Filling the reservoir is accomplished by closing a valve beneath the nipple (10), closing the valve (1), and venting the trapped pressure through another valve (2). Then the plug is removed (16) and filled. To put the unit in service, replace the plug, close valve 2, and open the valve beneath the nipple. The valve connected to the nipple is normally a 1- or 2-in. valve that is actually part of the wellhead manifolding or Christmas tree. Gradually open valve 1 and adjust to the drops per minute requirement as observed through the sight gauge (18). The bypass is for maintaining an equalized pressure across the reservoir in order that the rate of injection hopefully remains proportional to the liquid head in the reservoir. The unit must be properly braced and supported, but this is not shown because the technique must be compatible with the other associated facilities.

Note that all wells do not produce the same volumes except in a few instances. This may be a product of the type of reservoir, or it may be achieved with installed mechanical components. In either case, methanol injection can be greatly simplified by preparing an injection schedule based on the withdrawal rate and the number of wells in service. The schedule is posted on the work location bulletin board and, with a glance, the operator just coming on duty can tell exactly what quantities of methanol are required.

In spite of even the most elaborate and expensive precautions, occasionally there will be a freeze-off. When this happens, it often triggers a cascade effect and freeze-offs may begin all over the gathering system. Contingency plans should be made and executed

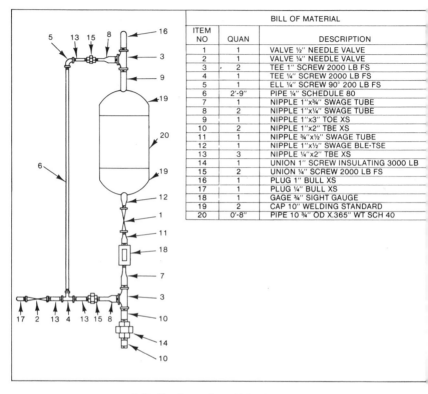

BILL OF MATERIAL		
ITEM NO	QUAN	DESCRIPTION
1	1	VALVE ½" NEEDLE VALVE
2	1	VALVE ¼" NEEDLE VALVE
3	2	TEE 1" SCREW 2000 LB FS
4	1	TEE ¼" SCREW 2000 LB FS
5	1	ELL ¼" SCREW 90° 200 LB FS
6	2'-9"	PIPE ¼" SCHEDULE 80
7	1	NIPPLE 1"x¾" SWAGE TUBE
8	2	NIPPLE 1"x¼" SWAGE TUBE
9	1	NIPPLE 1"x3" TOE XS
10	2	NIPPLE 1"x2" TBE XS
11	1	NIPPLE ¾"x½" SWAGE TUBE
12	1	NIPPLE 1"x½" SWAGE BLE-TSE
13	3	NIPPLE ¼"x2" TBE XS
14	1	UNION 1" SCREW INSULATING 3000 LB
15	2	UNION ¼" SCREW 2000 LB FS
16	1	PLUG 1" BULL XS
17	1	PLUG ¼" BULL XS
18	1	GAGE ¾" SIGHT GAUGE
19	2	CAP 10" WELDING STANDARD
20	0'-8"	PIPE 10 ¾" OD X.365" WT SCH 40

3–3   Ten-in. methanol injection unit

quickly in the event of a freeze-off because damage to the reservoir may result in certain kinds of gas storage fields.

Reservoir engineers say that the seasonal life of a storage field is based in part on the proportional deliverability of each well in the field. This consideration is more important to a true aquifer than to other reservoirs. Consequently, where freeze-off of a given well in a field is concerned at a near-maximum demand rate on the reservoir, it means that the other wells must overproduce to make up the deficit. In doing so, the latter wells will produce greater amounts of water, which in turn trigger a greater possibility of hydrate blockage in the gathering system. This will also make such wells water-off faster at the sand face and suffer a seasonal death. Thus, the importance of a scheduled methanol program and a continual pump

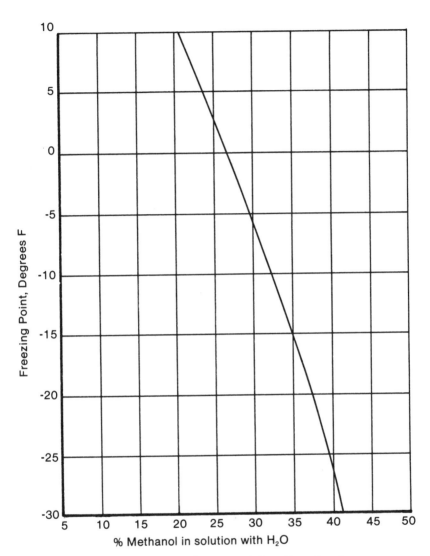

3–4 Freezing point of water in solution with methanol

maintenance and operating program cannot be overemphasized as the ultimate preventative measure.

There is one other source of blockage in at least some gathering systems that are designed with pipeline drips and some also with wellhead liquid-gas separators. These can be quite effective if they are installed at the proper location and are properly maintained. However, free water will accumulate in them and, even if equipped with automatic dump valves and liquid level controls (LLC), they are subject to freeze-off. One method to help prevent that from happening is to precharge the accumulator system with methanol. Fig. 3–4, which denotes the freezing point of water in solution with methanol, may be used to determine the quantity of methanol required for precharging these systems.

Another source of water accumulation is in the body cavity of certain types of valves, and these may also be precharged with methanol as an additional preventative measure.

Methanol, like any other chemical, is safe if properly handled and is noncorrosive except to lead and aluminum and alloys of each. It is toxic, and one should avoid inhaling the vapors, prolonged contact with the skin, and above all swallowing. It is flammable, has a relatively low ignition temperature (878°F) and should be stored as a hazardous material. Employees assigned to handle methanol should be knowledgeable and aware of the contents in "Chemical Safety Data Sheet SC-22," available from Manufacturing Chemists Association Inc., 1825 Connecticut Ave. NW, Washington, D.C. 20009.

PART IV DEHYDRATION

Thus far, water in both condensed and vapor forms has been a contaminant. Condensed water is in the liquid phase and occurs in the flowing natural gas stream in both slugs and minute particulate matter, the latter sometimes referred to as an aerosol. Entrained water is usually thought of in this manner. The liquid droplets are so fragmented and small that they move suspended in the gas stream due to velocity. Water present in this manner should not be confused with vapor. Water vapor is present in the gaseous state and occurs in the gas stream as a result of pressure/temperature equilibrium conditions acting upon the free or entrained water existing in the natural gas stream.

The same process occurs in the atmospheric environment. As the temperature increases, the rate of evaporation increases. When the atmosphere has absorbed as much water vapor as it can contain at a given temperature, the atmosphere is said to be saturated and the relative humidity is 100%. As the temperature declines, the vapor or gas condenses into liquid, and the result is rain or free water or entrained water.

This same cycle occurs in natural gas where water is present. The water in a gas pipeline or reservoir will occur as either a liquid, gas, or solid, depending upon the temperature. Because most pipelines are operated at elevated pressures, the latter is also a consideration because the quantity of water vapor that a gas can contain and the

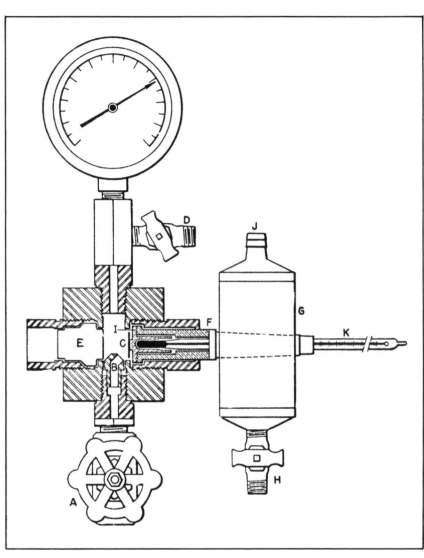

4–1 Sectional view of U.S. Bureau of Mines dew point apparatus *(courtesy U.S. Bureau of Mines)*

conditions for hydrates to form at constant temperature and volume vary with the pressure.

Dew point is a term frequently heard along the pipeline system and throughout production and storage field locations. As the atmospheric temperature declines, the water vapor tends to condense into a liquid or solid. The temperature at which this condensation starts to occur is the dew point. The lower the temperature declines, the greater the rate of condensation. This is exactly what occurs when cold water is poured into a glass and water starts to condense on the outside. If the temperature on the outer surface of the glass were measured and read at exactly the same time that condensed water began to appear, that temperature would be the atmospheric dew point.

This is the same process that occurs when a dew point of the flowing gas stream is determined, only it is accomplished with a sophisticated instrument designed to withstand high pressure. It is equipped with a pressure gauge, a highly polished steel mirror surface, a means of reducing the temperature of the mirror surface, and a thermometer to tell the exact temperature at which the water vapor starts to condense out of the gas as it is deflected onto the surface of the mirror (Fig 4-1).

Thus, dew point determination is specific to only the quantity of water vapor present in the gas stream and is only related to the amount or quantity of free water present. However, several assumptions can be made based upon the dew point test:

1. If the dew point occurred at a temperature less than the flowing gas pipeline temperature and at the same pressure, it can be reasonably assumed that water is only present in a vapor state—at least at that given instant.
2. If the dew point occurred at the same or at a higher temperature than pipeline temperature and at the same pressure, it can be most assuredly concluded that the gas is carrying entrained water in the condensed or liquid state, whether it be aerosol-sized or in slug quantities.
3. A gas is said to be saturated with water vapor when the flowing gas pipeline temperature and the dew point temperature are in equilibrium at constant pressure. Any time this occurs it may be concluded that there is also condensed water present in the flowing gas.

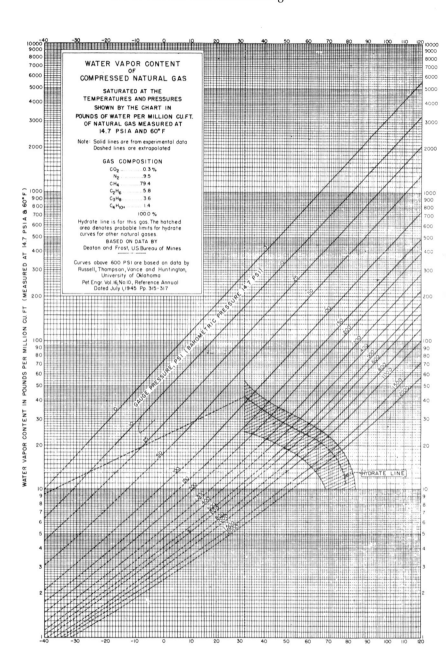

4–2 Water vapor content of compressed natural gas *(courtesy U.S. Bureau of Mines)*

Then why are water content and dew point important to the transmission of natural gas? Remember that small amounts of water could be considered diluents and larger quantities, contaminants. A pipeline flowing at 600 psi and 50°F with a dew point at 10° is operating at much less than pipeline saturated conditions and is carrying about 5 lbs of water vapor per million cubic feet, as shown in Fig. 4–2.

Certainly the quantity of water vapor present in this instance is of no consequential operating significance. The 600-psi gas could be expanded to 50 psi and no condensation would occur unless the temperature dropped to −30°F. This is the basis for a rather arbitrary industry-wide acceptance that gas containing 7 lbs of water vapor or less is considered okay. In this particular example, if the gas were expanded or regulated down to 50 psi, the resultant temperature would only be approximately 13°F above zero and still considerably above the 30°F below zero. Natural gas will reduce in temperature about 6.7°F for each 100 psi of pressure reduction, or (600# − 50#/100) (6.7) − (50°) = 13°. This cooling process in the presence of pressure reduction or expansion is known as the Joule-Thomson Effect.

Now, if this same gas indicated saturation at 600 psi and 50°, and recognizing a previous conclusion that if a dew point indicates saturation it can be assumed that free water is present, then a pressure reduction to only 300 psi would result in a 32-degree temperature. Water in condensed form will freeze at that temperature.

Even at 600 psi and 50° and saturated, the water vapor content is only 18 lb/MMcf. Since a million cubic feet of gas weigh 47,000 lbs, the percent of water in the total is only:

$$18/47{,}000 = 0.00038, \text{ or } 38/1000 \ (0.038\%)$$

But even the minimum effect could be disruption of service at some delivery point to a town, and that is quite serious.

Then there is the possibility of hydrate formation in the system. Hydrate stoppage most often occurs when a gas line is reported to have frozen-off. Technically, hydrates are compounds formed by the

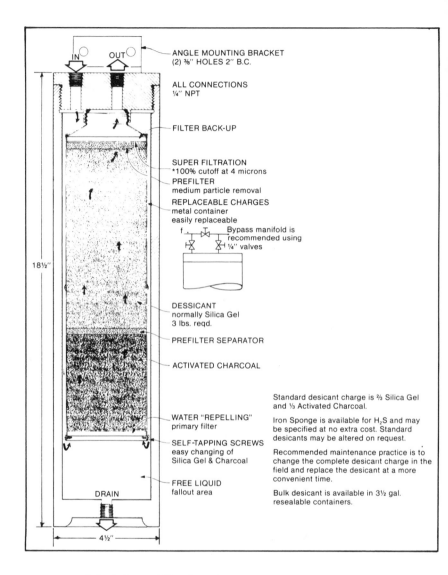

4–3 Instrument filter dryer *(courtesy Welker Engineering Co.)*

union of water with some other substance. Where natural gas is concerned, the substances are hydrocarbon gases. Numerous publications exist pertaining to hydrates and their formation. For those further interested, reference is made to the Bureau of Mines, Monograph 8, publication. Briefly described, a hydrate formation is normally crystalline in appearance, much like packed snow or chipped ice. Hydrate formation has been described as a phenomenon and occurs when the following conditions exist:

1. Presence of free water
2. Presence of hydrocarbons, principally methane
3. Turbulence of gas flow
4. Conducive pressure and temperature conditions

The water vapor content graph (Fig. 4–2) reveals that hydrates will form at temperatures considerably greater than the freezing point of water. The cross-hatched area indicates the probable conditions for hydrates to form. The range is quite broad and lies in the zone of the industry's general pressure/temperature operation. At 300 psig, hydrates can form in the temperature range of 40–55°F; at 500 psig, within the range of 48–62°F; and at 800 psig hydrates can form at temperatures up to 68°F. Turbulence, one of the conditions for forming hydrates, is the most prevalent at meter and regulator stations because of the increased velocities and the multiple and abrupt changes in direction of flow.

The graph also shows the lower the pressure at constant temperature, the greater the volume of water vapor possible in the gas. For example, a gas at 800 psig and 40°F could contain 11 lbs of water vapor/MMcf while at 100 psig the same gas could contain 52 lbs/MMcf.

A word of explanation at this point for those familiar with steam table data: Water vapor content is a product of the laws of partial pressure and temperature, but experimentation with compressed gases has proved that the water vapor content can be much greater than that theoretically calculated by use of such data. This is the reason for dehydration facilities at many storage locations and usually where gas is being taken from a producing field. Dehydration plants are designed to remove water vapor from the gas. Scrubbers and separators are designed to remove condensed water from the gas.

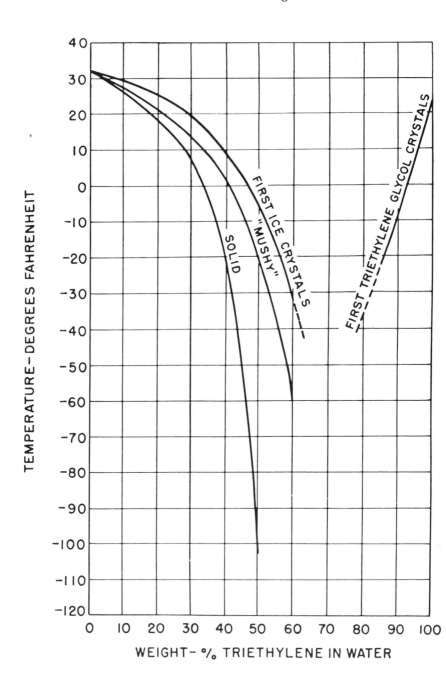

4–4 Characteristics of triethylene glycol-water mixtures at low temperatures *(after Polderman)*

There are several methods for removing water vapor from natural gas. Of those methods, two are in general use. The first and most predominantly used throughout the industry is the glycol system. The second utilizes solid compounds having a high affinity for water. The agents in these latter units are commonly referred to as *dry desiccants* and are produced by various manufacturers under trade names like Solva Bead and Mobil Bead. It can probably be identified with some of the various salt derivatives such as calcium chloride or absorbitives such as activated charcoal or alumina.

Another solid agent used frequently for small-volume drying is silica gel. Silica gel is most commonly used for drying instrument air supply. It is extremely good for this purpose because it is available as indicating. In other words, when it is dry, it is one color (usually blue); but as it becomes moisture-laden, it changes color (usually to pink then white). The Welker F4 and F5 dryers use this particular media. When it becomes saturated, it can be regenerated by baking in an oven or simply in a pan with heat applied. Where these units are installed, they should be checked as often as is possible until a timed experience factor can be established for regeneration purposes (Fig. 4–3).

As stated earlier, most large-volume dehydration units operating at storage and production locations are of the glycol type. The media is glycol or liquid absorbent, of which there are three kinds in general use: ethylene, diethylene, and triethylene. The latter, triethylene glycol (TEG), is the most frequently used for dehydration purposes. The technical reasons for using TEG instead of one of the other glycols are numerous and in some instances arbitrary. However, those considerations are beyond the scope and intent of this discussion. For this intent, it is accepted that TEG is the most satisfactory agent for general use because:

1. It has a high affinity for water
2. By comparison, the cost is reasonable
3. It is practically noncorrosive
4. It is a stable substance
5. Regeneration is continuous and simple
6. Compatible viscosity about 50°

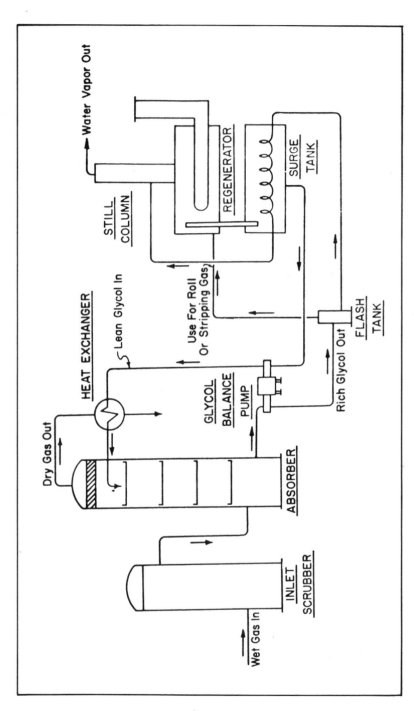

4–5 Dehydrator *(after Redus)*

7. The vapor pressure is low at operating temperatures
8. Low solubility for natural gases
9. By comparison, it has low foaming or emulsifying tendencies
10. Boiling point is much higher than the other glycols.

When water vapor in natural gas is absorbed by contact with TEG, it is described as hygroscopicity or the taking up and retention of water. We say that glycol has a high affinity for water or that water is soluble in glycol, like salt is soluble in water. Other than to recognize the action that takes place, it is of no particular consequence herein. That aspect is a part of design and specification rather than operating criterion. Of some interest may be that TEG will freeze, but adding water reduces the freezing point (Fig. 4–4).

The basic TEG dehydration facility normally consists of an inlet gas scrubber/separator, an absorber or contactor where the gas dehydration process actually occurs, a heat exchanger where the lean solution is cooled and the rich solution is preheated, sometimes a flash tank for venting absorbed gases, a filter system, a gas-fired boiler, sometimes with a stripping gas system, a still column with a reflux system, and a storage tank, making up the regeneration or redistillation unit. Interconnecting rich and lean solution piping locks the system together. Auxiliary equipment includes high-pressure displacement pumps, circulating pumps, liquid-level valves, a boiler fuel system, level controllers, and thermostatic controls. Figs. 4–5, 4–6, and 4–7 show parts of the facility.

The process and flow through the system is described as follows: The wet inlet gas enters the primary scrubber, where any liquid accumulations are removed. The separator may be classified as a two-phase unit: a vessel large enough that the velocity is reduced many times, enabling the particulate matter to simply gravity out, and a mist extractor. The gas then enters the bottom of the TEG-gas contactor. In most units, the contactor is a three-phase system whereby the gas enters an accumulator section, allowing any additional particulate matter not removed by the scrubber to also gravity out of the stream. Actually, any designation as to the number of

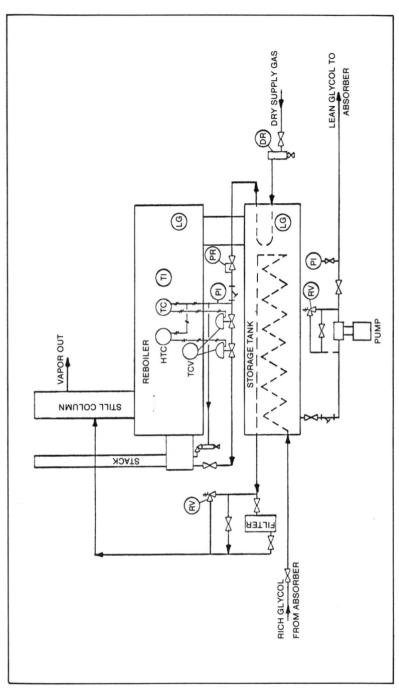

4–6 Glycol regenerator (courtesy Black, Sivalls, and Bryson)

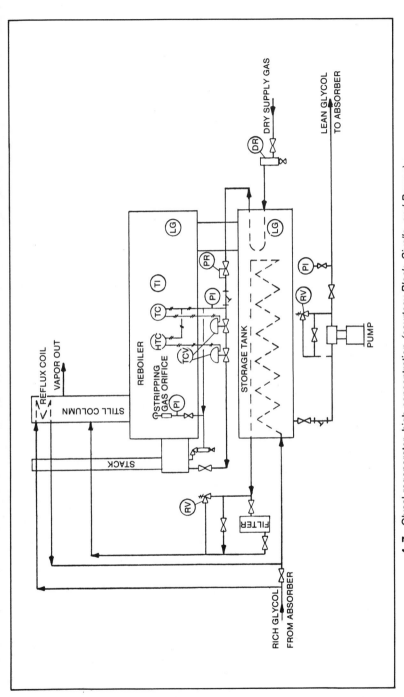

4-7 Glycol regenerator, high concentration *(courtesy Black, Sivalls, and Bryson)*

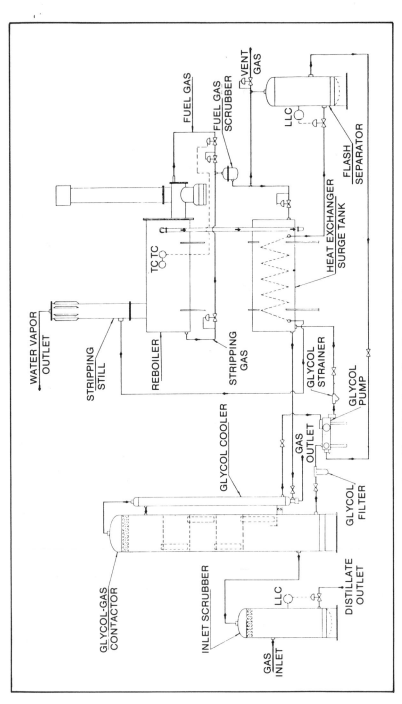

4–8 Glycol dehydrator flow diagram (after Sivalls)

phases that may be incorporated into an absorber (or, for that matter, into any part of a purification system) is a misnomer. Many auxiliary devices of known principle can be incorporated into a system if they are required by the user. For example, an additional heat exchanger may be incorporated into the unit, as shown by Fig. 4–8.

After the gas enters the bottom of the contactor or the scrubber or accumulator section, it flows upward through a mist extractor or, depending upon the manufacturer, through some type of chimney/coalescer, wherein hopefully the final quantities of any remaining free liquids are stripped from the gas. Where the TEG contactor is downstream of and in series with an MEA absorber, the liquids captured in the TEG phase-one scrubber section are manifolded into the MEA circulation system where they are reclaimed. The flowing gas then enters the bubble-tray sections where contact is made with the lean TEG. After passing through 4 or 5 bubble-tray sections, the gas then passes through 1 or more mist extractors where any entrained TEG is removed before reentering the pipeline system, supposedly as a dry and clean product. Design criterion determines the number of bubble-cap trays. Some systems utilize only 2 trays, whereas other manufacturers will use 4 or 5. Regardless of the number, the first tray removes the greater quantity of water vapor and the percent of removal declines to the top tray where its removal effectiveness may be only a fractional percentage of the first tray.

Fig. 4–9 is a graph showing the action at each tray for a six-tray contactor. Temperatures tend to average gas and solution temperatures at each tray. In some instances, the gas is further processed through an additional outlet scrubber as a final effort. Fig. 4–10 is a cutaway cross-sectional drawing in perspective that shows the internal makeup of a typical TEG contactor. Fig. 4–11 is another sketch of a contactor with a blow-up of one type of bubble cap. Again, and mostly depending upon the particular manufacturer, bubble caps are of different designs but each has the same ultimate functional objective.

As depicted by the figures, the lean TEG enters the top of the

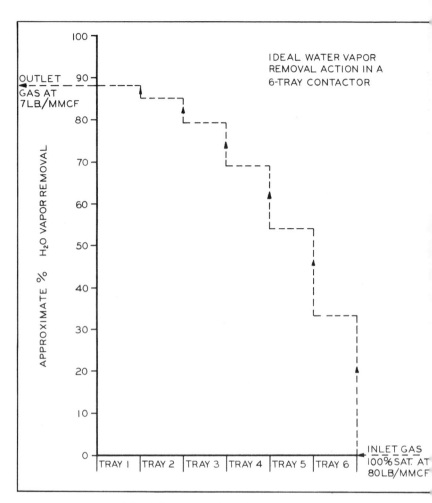

4–9 Action per tray for a six-tray contactor

contactor and gravitates downward through a series of downcomers, or chimneys, from tray to tray. Each tray is equipped with some type of weir arrangement for TEG retention on the tray at a level that requires the gas to make intimate contact. The rich TEG is then accumulated in a section equipped with an LLC (liquid level controller) that operates a control or dump valve. In some very low pressure applications, it would operate a pump. Usually, though, the contactor pressures are sufficient to force the rich TEG back into

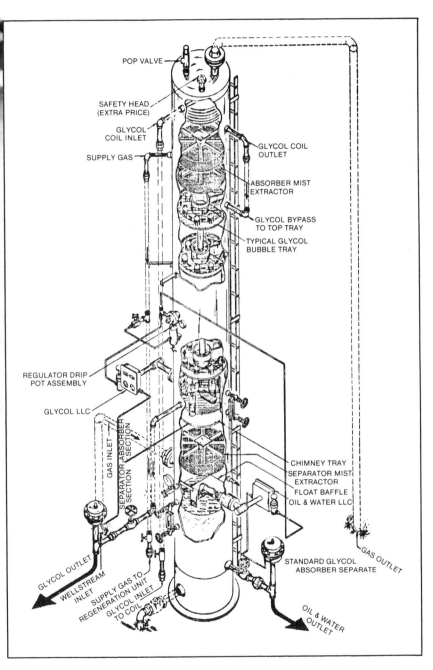

POP VALVE

SAFETY HEAD
(EXTRA PRICE)

GLYCOL
COIL INLET

SUPPLY GAS

GLYCOL COIL
OUTLET

ABSORBER MIST
EXTRACTOR

GLYCOL BYPASS
TO TOP TRAY

TYPICAL GLYCOL
BUBBLE TRAY

REGULATOR DRIP
POT ASSEMBLY

GLYCOL LLC

GAS INLET

SEPARATOR ABSORBER
SECTION SECTION

CHIMNEY TRAY

SEPARATOR MIST
EXTRACTOR

FLOAT BAFFLE

OIL & WATER LLC

GLYCOL OUTLET

WELLSTREAM
INLET

SUPPLY GAS TO
REGENERATION UNIT

GLYCOL INLET
TO COIL

STANDARD GLYCOL
ABSORBER SEPARATE

GAS OUTLET

OIL & WATER
OUTLET

4–10 Typical TEG contactor *(courtesy Black, Sivalls, and Bryson)*

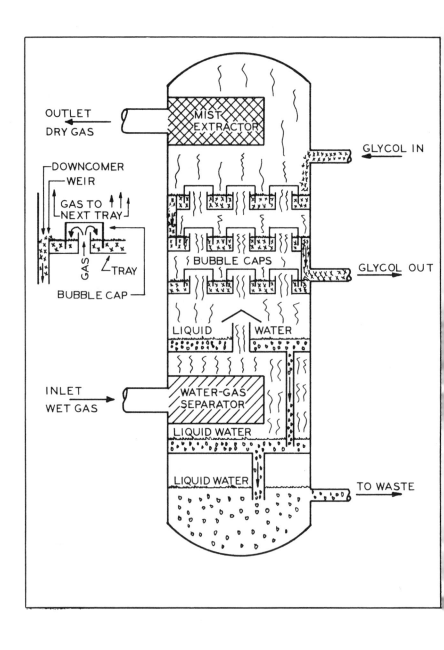

4–11 Gas-water-glycol flow through a contactor

the redistillation system. Positive displacement or plunger pumps force the lean TEG into the top of the absorber. A cliche to remember in reference to a solution-type gas conditioning system is simply that "lean is clean." Rich solution contains all the contaminants removed from the gas.

The rich solution, upon leaving the LLC valve, enters a filter for precleaning (not all systems use a filter). It then enters a flash tank where the absorbed gases are flashed-off to the atmosphere, used as fuel, or maybe used for stripping gas. The solution is then preheated before finally entering the still column and reboiler system for redistillation before again being recirculated through the continuous process cycle.

Some regenerative units are equipped with a stripping gas system. The stripping gas assists in reconcentrating the TEG to a higher percentage, thus increasing its affinity for water vapor and in turn increasing the efficiency of the system. The quantity of stripping gas to be used varies between 2–10 cu ft/gal of glycol circulated. Also, the still column is equipped with a reflux system, wherein the water vapor escapes to atmosphere but the glycol vapors are recondensed and gravity back into the circulation system.

The still column may be either a packed column or, in some instances, a tray column. Usually the column is packed and the materials normally used for this are of ceramic materials, called Berl saddles or Raschig rings. They vary in size from about ¼ inch to ½ inch, and the Raschig rings look like a finger handle on a coffee mug. Berl saddles are about the same dimension but are shaped somewhat like a saddle.

This packing is supported in the column by a screen between the tower and the reboiler. This screen should be stainless steel; however, older units have screens made of simple carbon steel. These have been known to corrode and dump the packing into the reboiler. Also in some units where the storage tank is located under the reboiler and connected with a downcomer and weir arrangement, this may be a packed section. Heavy contaminant loading of the solution will tend to plug the packing and cause many problems, one of which would be back pressure created on the kettle, in turn producing a higher boiling point and consequently less-efficient dehydration. The reflux system on glycol still columns is normally a fin-tube arrangement that tends to take advantage of atmospheric cooling and resultant condensation of glycol vapors. However, some columns also circulate the cool rich glycol through a coil to enhance the reflux action.

A glycol dehydration system that is properly designed and mechanically equipped will provide a minimum of operating problems if routinely maintained. There are, however, many considerations. While it is impossible to enumerate every possible and potential trouble spot, there are certain routine inspections, observations, and maintenance functions that, if rigidly adhered to, will tend to reduce major deviations from occurring.

During any period of continuous operation, it may be necessary to blow down an absorber or contactor or it may be shut-in for some reason. In either instance, it is likely that the pressure on the vessel will be less than normal operating pressure. Therefore, when bringing the vessel up to operating pressure, do so very gradually. Most of these units are designed for very low-pressure differential at capacity flow rates. Sudden pressure impacts will create excessive differentials across the internal parts—trays, mist extractors, etc.—and the result could wipe out or gut the unit.

When placing a system into operation, circulate the solution until the required operating temperatures are achieved before inducing any gas flow. Glycol is quite viscous, particularly at temperatures of 50°F or less, and gas flow at these temperatures will cause excessive glycol loss and carryover into the pipeline system. In so doing, additional contaminants are introduced into the system, negating the purpose for which the system exists. It is of utmost importance that solution flow be established before introducing gas into the system, and the gas must be induced very slowly. Many bizarre things can occur if done otherwise: foaming, solution overload of the upper trays and mist extractor, loss of glycol, and, more importantly, no dehydration occurs. Gas and solution-tray and bubble-cap interaction is a very carefully designed process, and the violent interplay that must happen will not occur unless the process is very carefully initiated.

Each location should have a flow diagram with optimum or manufacturer specified solution and vessel temperatures indicated. If the temperatures depart from those specified by more than a few degrees, process troubles are indicated and the cause should be found as soon as possible. The most critical temperature is the reboiler; it should maintain 400°F as nearly as possible. A point of caution: TEG begins to break down or decompose at temperatures about 400°F and above. Thus the maximum operating temperature must be 400°F.

Fig. 4–12 indicates the critical effect of reboiler temperatures. TEG redistilled at 400°F produces a 99% pure concentrate. Then, Fig. 4–13 shows that a 99% TEG concentration will produce a dew point depression of 64–78°F when circulating 3–12 gal/lb of water-vapor content.

Foaming of the glycol is another problem to be avoided and, in addition to system dehydration inefficiency, foaming of the solution greatly increases glycol loss (Fig. 4–14). The point of foam inhibitor injection into the system is a topic of much debate with various experts offering different opinions. The most effective point for injection is some place upstream of where the action seems to occur. In the absence of specific knowledge, a good place to start is upstream of the filter system, near the flash tank, or directly into the reboiler. Most foam inhibitors have a short life, and their inhibitive qualities will completely dissipate; therefore, the agent must be continuously injected into the system in small amounts with a chemical injector pump.

Antifoaming agents are available from practically all of the chemical companies. They must be used according to supplier instructions. This is particularly critical as to the quantity to be used. Injecting a greater amount than specified can reverse the hoped-for result, i.e., too much can create a foaming problem. One very important point to remember is that something, some impurity probably, is causing the foaming problem. Glycol being circulated too cold is one cause. Another is failure of the inlet separator. The best solution for foaming is to determine the cause and eliminate it. Not all defoamers will work in a particular system, depending upon the cause of foaming.

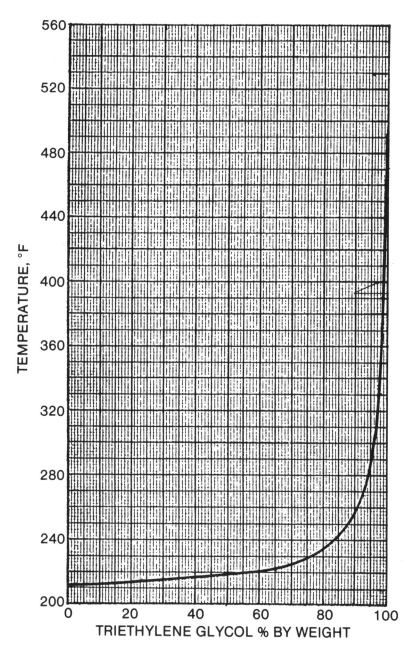

4–12 Triethylene glycol-water boiling point vs. composition, 760 mm *(courtesy Carbide and Carbon Chemical Co.)*

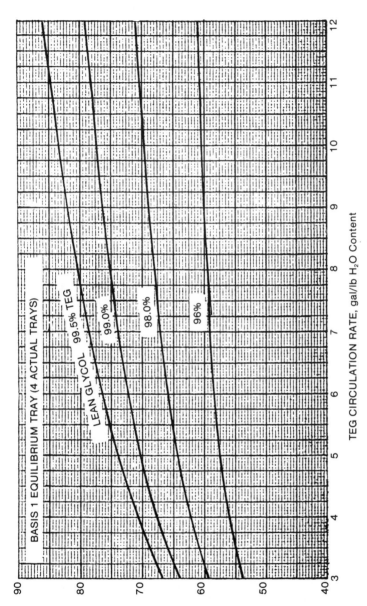

4–13 Dew point depression vs. TEG circulation rate *(courtesy Magnolia Petroleum Co.)*

LBS/MMcf

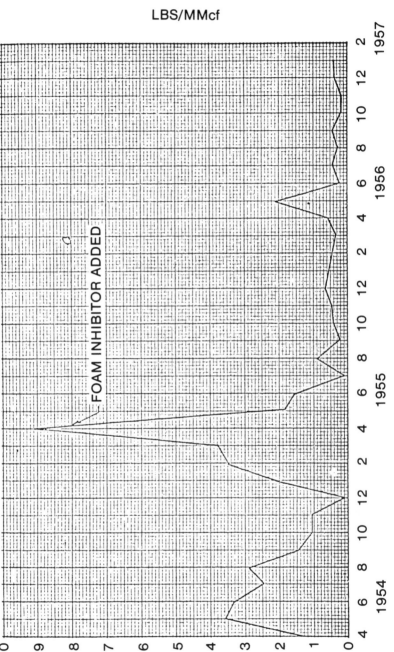

4-14 Effect of foam inhibitor on glycol loss *(courtesy University of Oklahoma)*

Another potential trouble spot is the main burner and fire-tube system. A proper flame is most important. Because burners are of many different designs, the manufacturers' recommendations for burner pressure, flame appearance, and air input are quite critical and should be followed exactly. Poor fire-tube and burner practices will lead to an excessive corrosion rate, rapid deposit and hot-spot buildup on the outer fire-tube wall, and increased fuel consumption. Be sure the flame is aimed directly down the fire-tube center and not deflected toward the walls. All reboilers have a BTU rating based upon orifice or nozzle size and pressure stipulation at the inlet. If no manual is available, the manufacturer should be contacted and requested to furnish the necessary data.

One rather quick check of burner performance can be made by installing a meter in the fuel system. For example, a 1,000,000 BTU/hr burner will consume 1,000 cubic feet of 1,000 BTU gas in one hour when firing at maximum rate. Fire tubes in even the best-designed system will build up carbon, and carbon is an excellent insulator, thereby reducing the effectiveness of even a properly adjusted burner. This carbon buildup must be periodically removed, and the best way is mechanical and with a swab. There are, however, chemicals available that can be pitched into the fire box, inducing a flame reaction that attempts to blow the carbon out the exhaust stack.

Pumps, particularly high-pressure plunger or displacement units, are high-maintenance items and need constant attention. Overly tight packing glands will produce scoring and shortened plunger life. The glands should have some leakage of glycol, and that quantity should be captured, measured, and returned to the system. Excessive leakage indicates a need for plunger replacement and/or seal damage. Plunger scoring indicates solid contaminants in the system, and filters, scrubbers, and separators should be inspected. The apex of the required glycol circulation rate is the pump, which may be check measured with a drum while pumping to atmosphere or observing liquid levels. Leaking valves will drastically reduce the efficiency of these pumps. The required glycol circulation rate is further influenced by the particular manufacturer's design of the absorber or contactor. Circulation rates can be varied with pinched valve bypass manifolding.

As previously stated, glycol becomes highly viscous when exposed to temperatures of 50°F and below. Therefore, circulation should be maintained on a dehydration system during the colder months and when gas flow is predicted to be somewhat intermittent. One of the most important temperature considerations is that of solution/gas contact temperature at the contactor. Ideally they should be in approximate equilibrium, with the TEG slightly higher (up to 10°F) than the gas temperature and again ideally somewhere around 70°F. Inlet gas temperature should be above 50°F because of the previously mentioned high viscosity characteristics of glycol at temperatures around 50°F or less.

Certainly, there is little that can be done about gas temperatures normally because of ground temperature or contactor temperatures when the gas has just been desulfurized through an MEA plant. Operating temperatures are a product of conditions presented to the manufacturer for design purpose. The critical point is to be sure that all possible is being done to ensure contact temperatures are as close as possible to those specified by the manufacturer. Also be sure that the lower (down to 50°F) the inlet gas temperature is, the greater the dew point depression. Reduction in lean glycol temperatures may be achieved by improvising with additional circulation piping between the reboiler and absorber.

Glycol losses are nominal for a system in proper operation and should average about one- or two-tenths of a gallon TEG for each million cubic feet of gas passing through the absorber. This would mean that a plant facility with an annual 3,320 MMcf storage gas withdrawal capacity should use a minimum of 332 gallons and a maximum of 664 gallons annually. Usage in excess of this indicates serious problems, such as inadequate circulation, foaming, overcapacity, and mechanical problems in the contactor or pumps. Usage less than 332 gallons may also indicate equally serious problems. Most systems are equipped with some type of spill-over control. If the device is a relief valve dumping excess pressure and solution back into the suction, leakage is an ever-present possible problem that should be checked often and routinely.

The term pH or pH control is frequently used and is associated with almost every type of closed-solution circulating system. It relates to the hydrogen-ion activity and is nothing more than a universal index to express the degree of acidity or alkalinity content of a system on a scale ranging from 0 to 14 with 7 indicating neutral, meaning neither acid nor alkaline. Zero to 7 indicates acidity and 7 to 14, alkalinity. A glycol system tends to become acid over a period of time, and the pH moves toward zero. What happens is that the TEG tends to oxidize and cause organic acids, primarily formic acid and to a lesser extent acetic acid. The effect of a low pH index is a high internal corrosion rate, occurring first in the hotter regions of the system (fire tubes, still column, etc.) and also possibly on the contactor trays. The pH can be brought back to 7.0 by adding borax, or soda ash. The materials should be added to the system gradually. Large quantities added at one time will raise the pH level greatly above 7 and generate foaming conditions. pH meters are available for continuous monitoring of the system (pH content from 7 to 8 is desirable).

Most companies, however, have an engineering research group, including a chemical laboratory and assigned personnel quite expert in pH control technology. System samples can be mailed to the lab for analysis, and that group can also supply indicating equipment for field use. Supervisory personnel responsible for dehydration plant operation are advised to communicate regularly with the lab experts for direction and on-site assistance in this matter.

Capacity is a key word in dehydration plant operation. Each plant is designed to operate and dry a certain volume of gas passing through it at a given pressure. One criterion for this capacity is contactor velocity. The velocity at any given pressure should be that at design conditions. For example, a plant designed for 3 MMcf/hr at 800 psia has a capacity considerably less than that at 400 psia. Using the 3 MMcf/hr at 800 psia and knowing the absorber (contactor) has a 48-in. diameter,

$$V_{FPS} = \frac{0.6Q}{PA}$$

where:

V = linear velocity, ft/sec, for a typical 0.60 sp gr gas
0.6 = constant
P = pressure, absolute
A = cross-sectional area, in.^{2} (πr^2)
Q = scfh (standard cubic feet/hour)

$$
\begin{aligned}
V_{FPS} &= 0.6\,(3{,}000{,}000)\,/800\,[3.14\,(24^2)] \\
&= 1{,}800{,}000/1{,}446{,}912 \\
&= 1.244
\end{aligned}
$$

Thus, the design velocity is for 1.244 ft/sec. How much could safely be dehydrated if only 400 psia were available, where X is the new volume?

$$
\begin{aligned}
1.244 &= 0.6\ X/400\,[3.14\,(24^2)] \\
X &= (1.244)\,(400)\,(3.14)\,(24^2)\,/0.6 \\
X &= 899979/0.6 \\
X &= 1{,}499{,}965
\end{aligned}
$$

Then, the optimum capacity at 400 psia is 1,499,965 scfh. This exercise did nothing more than present a simple formula for computing velocity of gas volume flowing through a conduit. The essential point is to remember that the design capacity of a dehydration facility decreases about proportional to the decrease in working pressure.

Minimum capacities are another caution to be observed. The consequences of processing a minimal volume are not quite as dire as the overcapacity condition because the gas will just not leave the contactor dry. The reason again is that of solution interaction and the lack of violent contact at the lower velocities. For most dehydration plants, the minimum is about 25% of maximum design at constant pressure.

Another point of precautionary operation is vent stoppage, either in the flash tank or reflux system. Such stoppage can and will induce a back pressure on the reboiler, thereby raising the boiling point of the TEG with consequent reduction in the purity of the lean glycol. By examining Fig. 4–14, it becomes readily apparent that a solution purity of 2 or 3% less than design will not provide proper dew point depression at constant circulation rates.

Solution operating levels are equally as important as operating temperatures. Low or excessive levels in a properly filled system indicate circulation problems that can occur due to LLC valve malfunction, stoppage in filters, bubble trays, downcomers, or vent lines. Gas surging through the contactor will affect levels. Fire tubes must be covered with solution to help prevent hot-spot corrosion. The inability to maintain solution levels indicates serious problems.

Excess still column and reboiler loading is a problem that may occur on very cold and windy days. Water is being condensed out in the reflux system and is returning to the reboiler through the still column along with the condensed glycol, and the reboiler cannot maintain operating temperature. This water should be leaving the top of the column in the form of vapor or steam, but it is being condensed because the reflux temperatures are too low. Temperatures in the reflux region should be 170–210°F. If this occurs in a system where rich solution is being routed to a coil in the stripper, then this flow should be cut off, allowing all rich solution to flow to the main input point on the column. On other systems it may be necessary to reduce the gas flow or simply to shut down the gas flow and let the solution continue to circulate and cook off the excess water.

In essence, a TEG dehydration system's primary requisite is to prevent any type of fouling of the equipment and solution. While some of the causes for fouling have been examined, there is no way to anticipate all of the problems that might occur. The best preventive measures are a well-planned operational program and an annual turn-around maintenance plan. Dew point determinations should be part of the operational routine—after all, the dew point is the final analysis of a plant's efficiency.

The annual maintenance plan should take place during a time when the facility can be shut down. The system should be drained and the glycol filtered and cleaned. A water and solvent solution should be circulated until the plant is cleaned. Pressure should be applied to the shell-and-tube heat exchangers and other parts of the system to find where leakage or corrosion may have occurred. The fire tube should be likewise checked. An internal inspection of the reboiler and contactor is also desirable, particularly if suspected problems have surfaced.

Upon completion of the annual maintenance, the glycol can be placed back into the system or stored until later. However, under either condition the entire plant and storage tanks should have a positive psig gas blanket maintained on them during the entire down or standby time. This will assist in preventing corrosion and solution oxidization. TEG stock should be replaced and the plant started up at least a week or two before it is necessary for continuous operation. Cleaning solvent specification should be that recommended by lab personnel. Many solvents will leave deposits that may cause foaming.

Normally, dehydration plants are downstream of desulfurization plants. Consequently, a dehydration plant so installed, even though designed for the condition, must operate at maximum efficiency. Consider that the gas from a water solution desulfurization plant is saturated at much higher temperatures than the normal ground temperatures and the result is much greater quantities of water vapor to be removed. Extra effort should be given to dehydration equipment operating in that sequence.

Safety precautions for dehydration plant operation include all safety practices for the handling of natural gas. The main precaution is awareness.

1.   Be aware that a high-pressure gas is on the opposite side of the control valve or pump from the low-pressure fluid system.

2. Be aware that water saturated with hydrocarbon combustibles is being flashed off at vent pipes and through the reflux vent, and that freezing these vents could produce a hazardous situation.

3. Be aware that all vents probably contain combustibles and should be kept well away from fire boxes.

4. Be aware that vent lines should not be manifolded together, e.g., flash tank and reflux. Stopping the single outlet could multiply the hazard.

5. Be aware that some contaminants have very low ignition points and vessels are to be opened up and purged very carefully. All gases, vapors, and vapor-causing substances are to be removed before an individual enters a vessel, and a positive program is necessary for totally disconnecting and isolating any vessel from any source of gas.

6. Avoid any prolonged periods of inspecting vessels from open manways, inspection covers, etc., before they have been thoroughly evacuated.

Glycols are not a hazardous material under current Department of Labor definitions. But they are a chemical compound, and the usual precautions for handling such materials should be observed in conformity to all gas industry safety requirements.

PART V DESULFURIZATION

A proviso of nearly all gas purchase and sales contracts stipulates that no gas is to be sold or otherwise delivered that contains greater than one grain of hydrogen sulfide (H_2S) nor greater than twenty grains total sulfur per 100 cubic feet. It is for that reason that desulfurization facilities are installed and operated at certain H_2S-producing underground storage and production fields. Natural gas containing these quantities of sulfur is called sour gas. The sulfur gas that is present is known as acid gas, as is carbon dioxide (CO_2). This reference to total sulfur is of no great concern except that it includes hydrogen sulfide. The other sulfur derivatives that normally exist in typical natural gas are usually of such fractional or trace proportions that they are completely insignificant.

Actually, the quantities of sulfur with which most storage reservoirs are contaminated, probably a maximum of 75–150 grains/Ccf, are relatively small by comparison with some H_2S-contaminated production gas supplies. Nevertheless, they are much greater than that which can be sold. Hydrogen sulfide content of the inlet gas to the desulfurization plants (Figs. 6–1—6–6) will probably seldom exceed 40–100 grains/Ccf. Then, to give some perspective to the quantity of sulfur that must be removed, consider that 7,000 grains of hydrogen sulfide equal 1 lb mass. Therefore, forty 7,000ths (40/7,000) is 0.0057 lbs. It has been shown that 100 cubic feet of natural gas weigh about 4.7 lbs, and 0.0057 lbs divided by 4.7 equals 0.0012, or that a gas containing 40 grains/Ccf is about 0.12% sulfur.

The contract requirement of 1 grain/Ccf is 1/7,000, or 0.00014 of a pound, and 0.00014/4.7 equals 0.000030. Stated another way, the final product cannot contain greater than 0.003% hydrogen sulfide.

All of this simply emphasizes that the sulfur-removing purification process must be precise, because the initial product contains only a fractional quantity of sulfur, and even that must be reduced 40 to 100 times.

The presence of sulfur in the gas is very easily determined by exposing a piece of filter paper saturated with a 5% solution of lead acetate to a sample of the gas from a blowoff. If the paper turns yellow or tends toward yellow or gray, sulfur is in the gas. This is qualitative, i.e., it shows sulfur is present, but it doesn't show how much. A test to determine how much is quantitative is much more difficult to run. One such quantitative test is the Tutweiler, using an iodine solution. Qualitative tests are normally a part of the routine field operating assignment, while quantitative test by titration may be the responsibility of lab or technical personnel. However, many experienced operating people have, through repetition, learned to index the lead acetate test and consequently arrive at a very close approximation of the quantity of sulfur present.

Among other devices available for sulfur tests are the colorimetric indicator tubes, calibrated photoelectric cells (this is just a more sophisticated lead acetate test), the flame photometric sulfur compound detector—basically a chromatographic test—and probably several others. Test method references are ASTM publications 1355, DIO 72-56, D 2725-70, and D 1945-64 for those further interested.

In the natural gas industry, sulfur content is commonly referenced in terms of grains per unit of volume. However, in the manufacturing and equipment supply companies, the quantitative reference is usually in parts per million (ppm). Fig. 5–1 will assist in converting grains/Ccf to parts per million and vice versa in both gas and air. For example, 40 grains/Ccf is equal to 740 ppm in air and 1,240 ppm in 0.60sp gr natural gas.

The reasons for removing H_2S from the gas are numerous, but among the most important are:

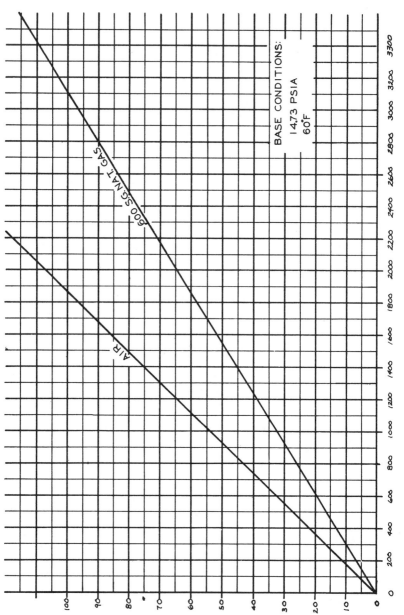

BASE CONDITIONS:
14.73 PSIA
60°F

600 SG NAT. GAS

AIR

PARTS PER MILLION

GRAINS/100 CUBIC FEET

5–1 Conversion table for gas and air, gr/Ccf to parts/MM

1. H_2S is corrosive to all metals normally in use and associated with pipeline systems, although some metals are less susceptible to corrosion than others. For example, it is assumed that H_2S is noncorrosive to stainless steel. While this is not quite fact, it is acceptable for general considerations. Brass and brass alloys are highly susceptible to H_2S corrosion, and they should not be exposed in any manner to gas containing H_2S.
2. H_2S is highly toxic.
3. H_2S forms SO_2 (sulfur dioxide) on combustion, which is also highly toxic and corrosive. (Solid deposits sometimes observed around burner outlets probably contain large quantities of SO_2.)
4. Both H_2S and SO_2 emit a foul odor like rotten eggs or a burning match.

Gas sweetening, or the removal of H_2S from otherwise sour gas, can be accomplished in a number of different ways. H_2S can be removed with as simple an operation as a plain water wash. In certain cases, this can be economically and practically feasible, but the method is not widely used.

Another method of H_2S removal and one that leaves the CO_2 in the gas is called the Iron Sponge process. The disadvantage of this is that it is a batch-type function and is not easily adapted to continuous operating cycles. The Iron Sponge is simply the process of passing the sour gas through a bed of wood chips that have been impregnated with a special hydrated form of iron oxide that has a high affinity for H_2S. Regeneration of the bed incurs excessive maintenance and operating costs, making this method inconsistent with an efficient operating program. If there are any real advantages in using this process, it is the fact that CO_2 remains in the gas, thereby reducing the shrinkage factor, which could be significant for very large volumes with an otherwise high CO_2 content. One company's average CO_2 content before desulfurization by MEA is about 1%, and the volumes processed at a maximum are about 3,500 MMcf annually. One percent of this volume is about 35,000 Mcf. Even at

Table 5-1
Toxicity of hydrogen sulfide to men

Hr S	0-2 Min	2-15 Min	15-30 Min	30 Min-1 Hr	1-4 Hr	4-8 Hr	8-48 Hr
0.005 0.010 50-100 ppm				Mild conjunctivitis; respiratory tract irritation.			
0.010 0.015 100-150 ppm		Coughing; Irritation of eyes; loss of sense of smell.	Disturbed respiration; pain in eyes; sleepiness.	Throat irritation.	Salivation and mucous discharge; sharp pain in eyes; coughing.	Increased symptoms.*	Hemorrhage and death.*
0.015 0.020 150-200 ppm		Loss of sense of smell.	Throat and eye irritation.	Throat and eye irritation.	Difficult breathing; blurred vision; light shy.	Serious irritating effect.*	Hemorrhage and death.
0.025 0.035 250-350 ppm		Irritation of eyes; loss of sense of smell.	Irritation of eyes.	Painful secretion of tears; weariness.	Light shy; nasal catarrh; pain in eyes; difficult breathing; conjunctivitis.	Hemorrhage and death.*	

0.035 0.015 350-450 ppm	Irritation of eyes; loss of sense of smell.	Difficult respiration; coughing; irritation of eyes.	Increased irritation of eyes and nasal tract; dull pain in head; weariness; light shy.	Dizziness; weakness; increased irritation; death.	Death.*
0.050 0.000 500-600 ppm	Coughing; collapse and unconsciousness.*	Respiratory disturbances; Irritation of eyes; collapse.*	Serious eye irritation; light shy; palpitation of heart; a few cases of death.	Severe pain in eyes and head; dizziness; trembling of extremities; great weakness and death.*	
0.060 0.070 0.080 0.10 0.15 600-1,500 ppm	Collapse;* unconsciousness;* death.	Collapse;* unconsciousness;* death.*			

*Data secured from experiments on dogs which have susceptibility similar to men.
Source: National Safety Council data sheet D-chem. 16

$2/Mcf, this amounts to only $70,000 annually in cost for shrinkage. This would not begin to equal the additional cost of operating an Iron Sponge at those locations. Now if the CO_2 content were 10 or 15% and the gas were still a 1,000-BTU value with the CO_2 in it, the result would probably warrant at least further investigation.

This is only to recognize that there are other methods for removing H_2S. Most companies, nevertheless, utilize the amine process, i.e., a solution of 15–20% by weight monoethanolamine (MEA) in water. A recent survey (1980) conducted by an international engineering firm concluded that three-fourths of all gas desulfurization facilities in North America used the MEA or DEA (diethanolamine) process. There are even some plants designed to use TEG and DEA in solution for simultaneous dehydration and desulfurization.

MEA is the preferred media rather than DEA because it enables a smaller circulation rate of solution, although it has a higher vapor pressure with increased chance of higher chemical losses. The detrimental aspect of an MEA plant is that it is subject to expensive corrosion problems, suffers expensive high solution loss due to vaporization, and removes CO_2 that could better remain in the gas.

The MEA physical plant and its hardware are much like a dehydration system. The main differences are that the contactor contains many more tray sections, routinely referred to as bubble-cap trays. However, some MEA contactors use a valve-type tray whose function is identical to the bubble cap but is a working rather than stationary design. Not uncommon is a 20 or more tray contactor.

The more efficient MEA systems include a side-stream reclaimer and normally a more sophisticated filtering and heat exchanger application than does the TEG dehydration system. Operating temperatures are decidedly different, and the reflux and still column are larger and more complex.

Amine plant operation is inclusive of all of the precautionary considerations noted for a dehydration facility. There are, however, many additional considerations, all of which are far more critical and demand greater adherence to daily operational and maintenance standards. All of the problems encountered in dehydration operation are of much graver consequence when occurring in the amine system, i.e., foaming, corrosion, solution loading, heat exchanger fouling, solution loss, poor solution filtering, inlet gas contamination, and plant start-up, to name only a few. Additional problem areas in the amine plant operation are the product of the high vapor pressure characteristic of amine and resultant boiling point, particularly in the presence of water. Solution loading is critical, and the results are excessive heats of reaction in the contactor and excessive vapor pressures in the still column and regenerative system. Normal corrosion problems are further enhanced or intensified by the susceptibility of stress corrosion. Freezing of auxiliary devices does occur and can create tremendous problems, one of which is also loss of solution.

Theoretical MEA losses are in the range of 1.33 lbs/MMcf processed (9.0 lbs equals one gallon), or roughly 1.5 gal/10MMcf, and this includes a small amount for mechanical losses. Usage greatly exceeding these amounts indicates either or both serious mechanical and housekeeping problems. For continuous operation, a well-designed system with a reclaimer and under high standards of maintenance will require about 130 gallons of water makeup for each gallon of MEA makeup.

MEA has a high capacity for absorbing the acid gases (H_2S and CO_2) through chemical reaction, and this occurs in the contactor or absorber. That reaction is reversible at a higher temperature in the stripper (still column and reflux) and reboiler regenerative system. This permits reusing the original solution and provides the required continuous operating cycle.

The required circulation rate in gallons of MEA solution per minute is a product of design based upon the physical conditions of the gas and the end result expected. For general operating purposes, it can be stated that one gallon of 15 wt% solution will absorb approximately 2.5 to 3.5 standard cubic feet of acid gas—H_2S and CO_2. Thus, it follows that one gallon of 20% MEA solution (a typical design standard) will absorb about 3.5 to 4.5 scf of acid gas, or an average of 4.0 scf. Then, if it is required to treat 30 MMcf/day of gas

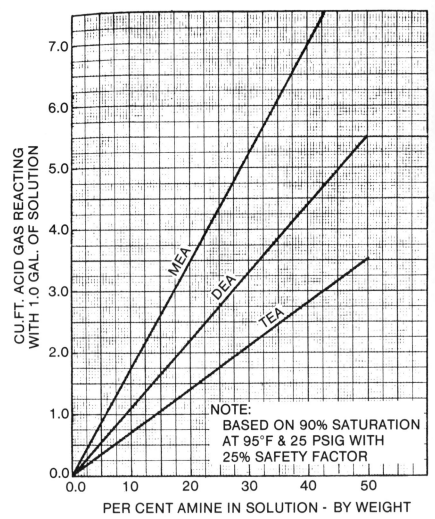

5–2 Acid gas pick-up with varying strength amine solutions *(after Connors)*

containing 1% CO_2 and 0.12% H_2S, what is the gpm circulation rate required for a 20% wt MEA solution?

$$GPM = \frac{(MMcf/d \; (CO_2 + H_2S)) \, / \, (scf/gal)}{(min/day)}$$

$$= \frac{(30,000,000) \, (0.0100 + 0.0012) \, /4.0}{1,440}$$

$$= 58$$

Thus, a 20% MEA solution rate of 58 gallons per minute is necessary. Obviously, this must vary slightly with solution strength at any given time, but it does provide the operator with a means of checking. If there is a sudden decline in outlet gas quality, it could be because the circulation rate is insufficient for the solution strength, or possibly because of a sudden increase of inlet acid gas, that being unlikely. Note also that solution strengths are in terms of percent by weight rather than percent by volume. MEA is about 8% heavier by volume than water when both are at the same temperature. In other words, when mixing solutions by volume, i.e., gallons per gallons, use about 8% more water. For example, when mixing a 20% solution, use 20 gallons MEA and 86, not 80, gallons of water. Thus, maintaining manufacturer specified circulation rates is very important.

Fig. 5–3 is a flow diagram of a typical MEA acid gas removal system. Each plant as fabricated, specified, and designed by individual manufacturers will often deviate slightly from that which is shown as typical, i.e. specific location of the filtering system and flash tank. Some systems include the side-stream reclaimer, while others do not. Some provide automatic solution makeup using on-stream solution storage reservoirs. Some utilize packed columns, while others use bubble-cap tray assemblies, particularly in the stripper column. And some have a very sophisticated reflux system. Other plants use fan-type aerial coolers or a combination of both fan and two- or three-pass tube and shell units. Many of these considerations are products of user specified requirements and initial costs. The reclaimer system is an example of the latter. It can be deleted but its incorporation is a question of future operating and maintenance cost.

Most of these considerations as to need and location in the solution circuit are the result of manufacturer design. As important as those considerations are, they do not directly apply to this effort other than where they become a vital part of the operating and maintenance requirement. Those considerations are built into the plants as constructed, whose operation must follow accordingly. Thus, the flow diagram is designated as typical rather than specific.

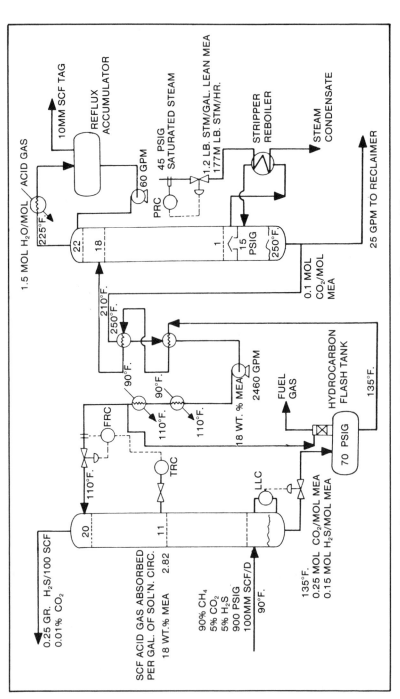

5-3 Monoethanolamine in the Girbitol process (*after Dingman and Moore*)

Each work location should have a flow diagram of the plant; if not, one should be prepared and prominently displayed for operating purposes.

The lean MEA is pumped into the top of the absorber and flows downward from tray to tray, making countercurrent contact with the sour gas entering near the bottom. The sweetened gas leaves the contactor at the top after passing through about 20 or more tray assemblies and a mist extractor. Many systems include an outlet gas scrubber where carryovers are further captured and returned to the system (Fig. 5–4).

The amine reacts chemically with the acid gas, and the rich solution is discharged from the bottom of the contactor and starts on its way through the regenerative cycle. The heat-of-reaction effect that occurs because of the intimate contact between the MEA and the acid gases in the contactor is shown in Fig. 5–5. Essentially, the temperature increases as the solution passes downward toward the accumulator section. This temperature increase can be quite severe, depending upon the percent of acid gases present in the stream. In this example, the inlet lean MEA temperature is only 100°F and the inlet gas temperature only 90°F, yet the maximum contactor temperature is around 240°F at about the third tray. About 90% of the acid gases are removed in the first three or four trays, thus the cause for the indicated increase in heat-of-reaction temperatures. This example is for rather large percentages of acid gas, but it does illustrate what occurs and that it is proportional to the acid gas content.

After leaving the contactor, the rich solution enters a flash tank that allows the absorbed gas contained in the solution to be vented. Retention times in the flash tank are variable, but as a general rule two minutes retention time is considered sufficient. In other words, if the circulation rate is 60 gpm, then the flash tank should have a minimum capacity of 180 gallons. Note that this flow diagram routes the flashed gas back into the fuel gas system. Another purpose of the flash tank is for a sediment accumulator. Obviously, the sediment or sludge must have provision for removal and not be allowed to accumulate continuously. Not all systems have a flash tank, this

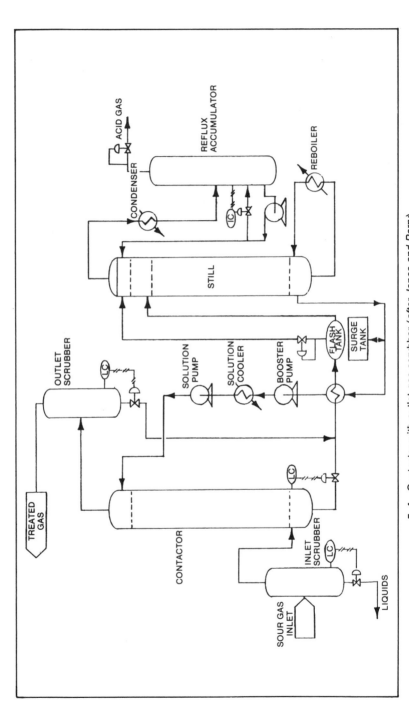

5–4 Contactor with outlet gas scrubber (*after Jones and Perry*)

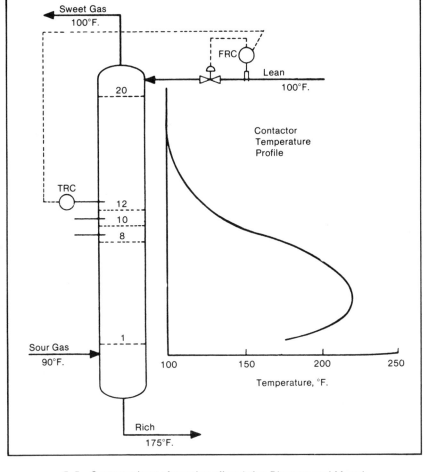

5–5 Contactor heat-of-reaction effect (*after Dingman and Moore*)

requirement being fulfilled in the reflux process. Some systems incorporate the flashing requirement into the filtering system. Where the amine contactor is upstream of and in series with a dehydration contactor, the inlet-gas liquid accumulator section of the dehydration unit may be manifolded into the rich MEA solution piping before the latter reaches the flash tank. This is to reclaim any MEA that was carried over into the sweetened outlet gas.

The location of the filtering system in the solution flow circuit is critical to the plant operation, as is the decision whether full-flow or side-stream filtration is used. In many instances, the initial design must be altered to meet conditions encountered through operating experience. Each operating group should review the past year's operating experience previous to construction budget submission time and request funds for needed changes or even additions. Operating characteristics and conditions of a particular storage or production reservoir can change from the original design criteria and will in turn require physical change in plant equipment. One consideration extremely important to the filtration system is the use of inhibitor materials injected into storage wells during the storage injection season. If the treating system is incurring excessive operating and maintenance problems, the operators should not simply "rock along" and assume that the plant cannot perform better. In many instances remedial action is available, while in others possibly nothing can be done or the capital costs are excessive in relation to the improvement that can be had.

Regardless, problems should be brought to attention for budgetary consideration and engineering study. As was previously pointed out, what may appear to be a filtering problem may in fact not be the fault of the filtering system but may possibly be caused by inefficient inlet gas cleaning and scrubbing, thereby requiring change or alteration of that system rather than the filtering system. Efficient filtration will provide reduced foaming, corrosion, and fouling of mechanical equipment, to name only a few of the benefits derived, and the degree of reduction accomplished is directly proportional to the effectiveness of the filtration.

Assuming that the particulate matter requiring filtration is reasonably constant in the solution, then the frequency of filter change is directly related to the quantity of solution circulated. This will in turn help to decide the type filters to be used because of replacement costs. This can be excessive, especially where traces of certain kinds of well-corrosion inhibitors are present. Certainly the disposable element sock-type filter is the most convenient, but its use could be economically prohibitive. With this type filter, costs could be as high as $5/1,000 gallons filtered. Obviously, in the previous circulation problem where the rate required was 58 gal/

min, that would necessitate filter replacement about every 17 minutes, or 1,000/58.

Optimum inlet separator and demister pad efficiency can minimize the problem of liquid and corrosion inhibitor carryover into the MEA system if the gas volume velocities are near the design values. But storage field withdrawal volumes may vary from as low as 5% to 100% of capacity. Thus, the velocity, as previously explained, will vary equally; consequently, also as previously noted, the components of separation are not effective. This is a problem over which the operator has little control. Nevertheless, he needs to be aware of it and offer recommendations for minimized withdrawal swings as well as the need for equipment change or alteration.

Where filter problems of this type exist, there is no real difference found between the types of sock filters available. Nor does its particle size removal rating make much difference, i.e., whether it be 5 microns or 25 microns, because a tight complex film develops on the element surface leading to a high pressure drop with no penetration depth of the solids it is supposed to capture. Throughput then becomes time dependent rather than solids-loading dependent.

One possible means of improving filtration problems resulting from entrained liquid and inhibitor carryover is to install diatomaceous earth (DE)—sometimes called Fuller's earth—filters. These filters involve a higher investment or installation cost but may achieve a lower operating cost. Iron sulfide is the most prevalent solid in MEA systems, and it's particulate size passes through 5-micron filter elements. But DE will remove particulate matter of one or so microns. DE filtration systems have been successfully used in the gas industry for many years. One frequent application has been in gas engine/compressor lube-oil systems. DE should be considered as a possible cure where a complex filtration problem exists.

The most desirable type of filtration is full-flow, i.e., the total solution stream is filtered whereas side stream filtration only involves a part of the full stream.

In summary, here are some operating considerations for MEA filtering systems:

1. Determine the differential at which the solution begins to bypass the filters and use this religiously to determine need for filter change.
2. Be sure inlet gas separating devices are operating at their maximum ability.
3. Do not allow solution to bypass filters for even a short duration. The system can become contaminated in a minimum time.
4. Install parallel secondary filters for item 3 or try to shut down long enough to change the filters.
5. Endeavor to operate as continuously near design capacity as possible.
6. Where the problem is overly severe and filter costs are excessively high, request an engineering study and recommendation toward filter system replacement.
7. Never start gas flow through the absorber until solution flow and temperatures have been established and only then.
8. Have used filters analyzed to determine depth or penetration of solids-loading. This will help reveal the effectiveness of the particular rating filter required.

The solution then passes through a heat exchanger system where it is further preheated by the lean solution flowing toward the contactor.

Next, the rich solution enters near the top of the still column/ stripper system where it gravitates downward, making intimate contact with the steam passing upward through the column as generated by the reboiler. This is the reversed action previously referred to where the steam strips the absorbed acid gases from the rich solution. The steam then passes through a condensing unit (oftentimes a fin fan) where condensed MEA liquid gravitates to the reflux accumulator and the acid gasses released in the condenser pass off to the flare gas and burn pit area—environmental considerations notwithstanding. (While environmental considerations are not a topic herein, the requirements exist and must be complied with accordingly.) The liquids accumulated in the reflux cycle are continuously circulated by pump and reinjected into the regenerative system. The lean amine accumulated at the bottom of the stripper column is constantly being recirculated through the reboiler, where hot vapors are simultaneously entering the stripper column from the reboiler. The lean amine is taken from the reservoir as required for return to the contactor, providing the suction for the high pressure pumps.

The temperature profile and heat-of-reaction in the stripper is not unlike that of the contactor. Fig. 5–6 illustrates that action.

Reboiler and flame attention is the same as required for a glycol system. The flame should be routinely inspected and carbon build-up in the fire box and stack should be removed. Flame characteristics should be as stipulated by the manufacturer for the particular burner employed. Of more current and future importance, as compared to past practices, conservation of fuel takes on greater dimensions. If there is any concern about a particular burner and its efficiency, an analysis of flue or stack gas exhaust should be made. There are several instruments available for this purpose, including the Orsat apparatus, made by several different manufacturers, and the Flue Gas Analyzer, made by Teledyne Analytical Instruments. If there appears to be a need for this test for properly adjusted burners, the operator should request assistance from the engineering group or other technical services personnel.

The next major component of an MEA system is the side-stream reclaimer (Figs. 5–7, 8). Fig. 5–7 schematically illustrates the position of the reclaimer in the solution flow circuit and the manner in which it is manifolded for inlet lean solution, outlet vapor or reclaimed solution in vapor state, and water injection for steam generation and cleaning in the reclaimer. The side-stream designation is because only about 1–3% of the lean 20% MEA solution flow is diverted to the reclaimer. If a system has an 8,000-gal capacity and

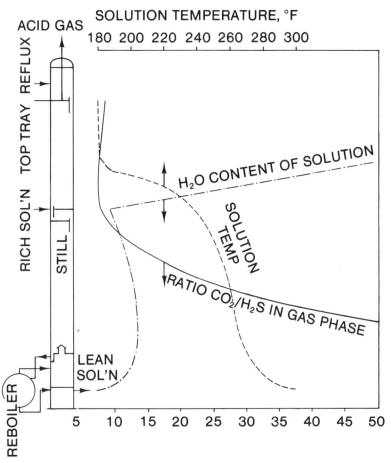

5–6 Still and reboiler temperatures and compositions in a plant *(courtesy* Gas Engineers' Handbook, *Industrial Press, 1977)*

the circulation rate is 60 gpm or 3,600 gph, then 36 gph is being processed through the reclaimer. In about 225 hours, 100% of the solution will have been through the reclaimer (8,000/36 = 225).

The reclaimer is nothing more than a smaller reboiler wherein the 20% lean solution is increased to an approximate 78% lean solution by boiling off the water, other impurities, and some of the amine in vapor state. These vapors are manifolded into the main

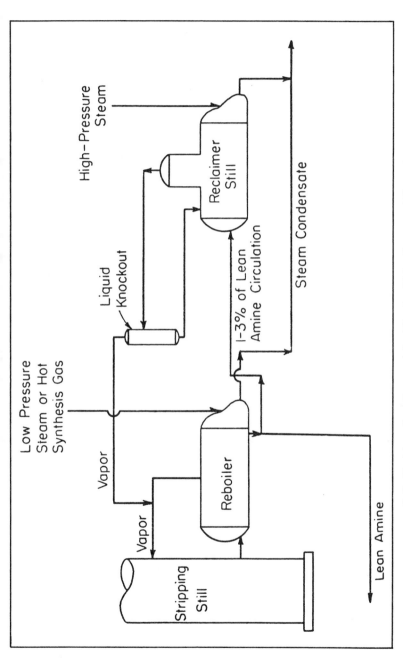

5-7 Reclaimer flow diagram (*after Wonder, Blake, Fager, and Tierney*)

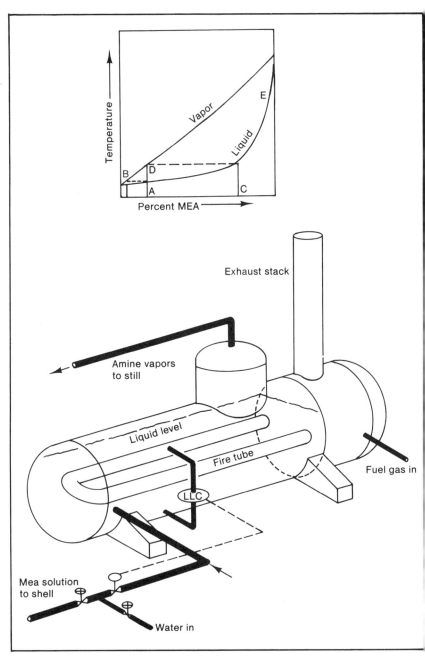

5–8 MEA vapor-liquid relationship and a typical gas-fired reclaimer *(after Shell, courtesy the* Oil & Gas Journal)

regenerative system still column where the amine is condensed and returned to the system. The water vapor and other impurities are flashed off to atmosphere along with the other products accumulated in the still from the rich solution. The sediment that enters the reclaimer settles to the bottom of the kettle and eventually must be flushed out with water and probably manually removed by pulling the fire tubes.

The hardware and instrumentation associated with a reclaimer system vary with the manufacturer and the requirements specified by the user. Following are items, some of which are needed for any kind of reclaimer operation, and others that may or may not be an existing part of the system. Items noted that are not a part of an existing system should be considered as necessary add-ons.

1. An LLC valve and controller.
2. An immersion thermostat control, sensing liquid bath temperature in the kettle and set to modulate the main burner for a lean inlet 20% solution at a control point of 255°F if kettle pressure is 0 psig; 263°F if 5 psig; 275°F if 10 psig.
3. A pressure gauge in the vapor section of the kettle or in the vapor vent line near the kettle.
4. A high pressure shutdown control if the above vapor pressure reaches a preset level. For example, a system where the still column is only slightly above atmosphere when in operation should have the pressure shutdown set at around 10 psig. This shutdown should close the fuel gas automatic modulating valve.
5. A relief valve set to dump back into the reboiler system at the above high pressure.
6. An indicating temperature thermometer in the vapor section of the kettle or close to the kettle in the vapor line. (Vapor temperature will usually be about 5–10 degrees less than liquid temperature.) This could also be an override temperature controller to shut down the main burner. This should be set at about 290°F and should also be a manual reset type.

This could even serve as a back-up override and could simultaneously close the LLC valve.

7. All input and output piping should have manual valves installed. The lean MEA and water input manual valves should be installed upstream of the automatic LLC valve.

8. The manual valve upstream of the fuel gas automatic modulating valve should be of a type that can be indexed so that the in-service process can be expediently repeated.

Satisfactory operation of the reclaimer system is vital to the rest of the system. Any malfunction can provoke serious operating problems in the main system, thereby making it most desirable that the reclaimer be isolated from the rest of the system.

The procedure for initially placing the reclaimer system in service and for continuous operation thereafter is as follows:

1. Be sure the reclaimer is empty.

2. Be sure the manual water valve, inlet solution valve, and fuel valve are closed, as well as any drain valves.

3. Set all controllers.

    (a) Modulate the fuel valve at 255°F, assuming it has a scale or set point and is properly calibrated. If not equipped with this type of control, set it toward open position and fine adjust after process is started.

    (b) Set all overrides to desired shutdown points.

    (c) Make sure LLC control valve is in open position, but do not activate with controller yet.

4. Fill reclaimer with 20% lean solution with manual solution valve from main reboiler to a level at least 5 or 6 inches above the fire tubes. Observe thermometers for 230 to 240°F.

5. Close manual solution input valve.

6. Activate LLC valve and controller.

7. Adjust LLC valve and controller to be sure they are in closed position and the level is 5–6 in. above the fire tube.

8. Open manual solution input valve.

9. Drain off enough solution to activate the LLC controller and observe it and the valve to ensure that it will maintain the proper liquid level in actual operation. Adjust if necessary.

10. Close manual solution control valve.

11. Be sure vapor line is open to still column.

12. Light burner pilot.

13. Open manual fuel valve to burner and adjust flame and air mixing unit.

14. Observe the temperature. It should rise to about 250°F in about ½ hour. If not, there must be a reason: insufficient fuel for ambient conditions, etc. Proper action is judgment.

15. If temperature does appear to be rising satisfactorily *open the manual solution control valve, immediately*. Any further delay at this point may tend to overload the still column with vapor, induce foaming, and cause the solution to fall below the required kettle level.

16. The system is now in automatic operation and must be further observed for the following conditions:
    (a) Temperature should continue to rise to 255°F.
    (b) The main burner should begin to modulate and maintain temperature around 255°F.

17. If the system does not reach the 255° temperature, it may be because:
    (a) The solution is foaming.
    (b) The thermostat is not quite adjusted properly.
    (c) The LLC units are not working properly.
    (d) There is a material buildup on the fire tubes—most likely on the outside of the tube in the kettle. Also possible is the buildup of carbon on the fire side of the tube.

18. Assuming all to be correct and as described, the system is in operation and will continue so if no solution nor mechanical problems occur.

19. Operating experience will determine when the system needs cleaning. The need for cleaning will be brought to

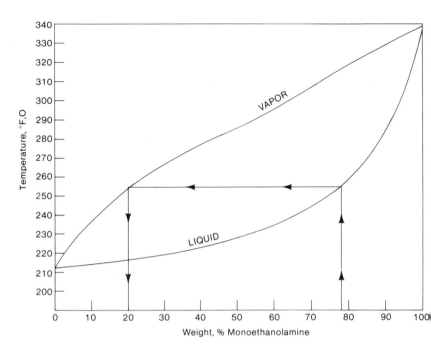

5–9 Vapor-liquid composition boiling point curve for aqueous MEA solutions (*after Conners, Shank, and Girdler Corp.*).

attention by observing departures from proper operating conditions described above, e.g., increasing and erratic temperatures, unstable liquid levels, foaming, change in vapor pressure. When this occurs, the lean solution feed valve should be closed and the water valve opened. The temperature process and consequent increase described above will begin to reverse and decrease to the boiling point of water (212°F) and will be attempting to reclaim the amine contained in the sediment. When the liquid temperature decreases to about 215°F, the entire system should be shut down and the reclaimer flushed out and possibly cleaned manually.

20. To replace in service, repeat the same process.

21. The foregoing assumes the system to be 20% operating solution strength and that the vapor pressure above the liquid is 0 psig. Under these conditions, a liquid tempera-

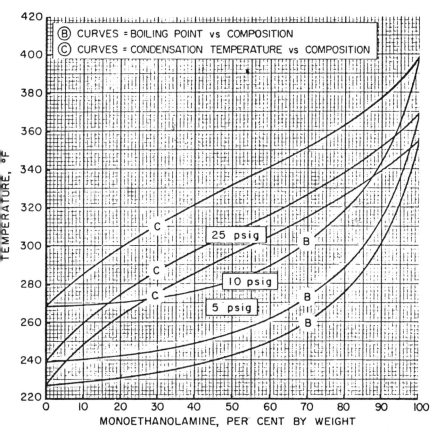

5-10 Monoethanolamine-water boiling point and condensation vs. composition at various pressures *(after Wonder, Blake, Fager, and Tierney)*

ture of 255°F degrees will produce a vapor content of 20% MEA and a liquid concentration of 78% MEA. The system is said to be in equilibrium.

In further explanation of the process, it is pointed out that the temperature increase in the initial 20% solution *is not* the result of the amount of fuel gas being supplied to the burner. Rather, it is because of the progressive increase in concentration of MEA in the kettle and the consequent increase in boiling point. The time that it takes to reach boiling point *is* a product of the burner flame. Figs. 5–9 and 5–10 illustrate the boiling point of MEA at atmospheric

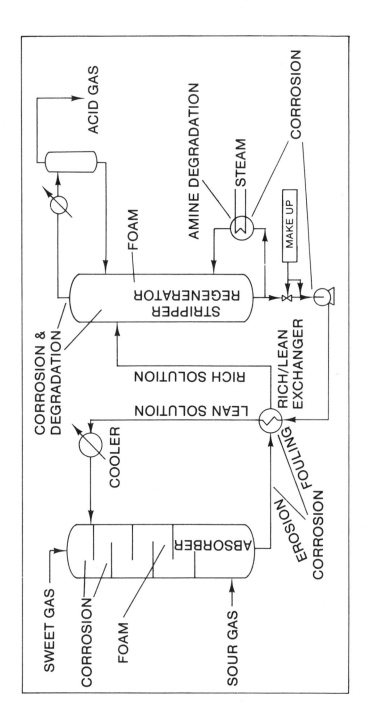

5-11   Corrosion, foam, and amine-loss sites *(courtesy the* Oil & Gas Journal*)*

pressure and, in Fig. 5–10, at 5, 10, and 25 psig. Note that the boiling point of 100% MEA is at 340°F, while at the other end the curve reflects the boiling point of water at 212°F. The requirement herein is for a 78% liquid concentration in the kettle. This means that the vapor being boiled off at the 255°F temperature is 20% MEA, and this is the volume at any given instant being returned to the system. Further examination of these curves will very readily explain why still column loading occurs if solutions are brought up to boiling point too rapidly. Summarily, any increase in heat input over and above that required to maintain the boiling point for any given concentration is wasted.

Foaming can be caused by too fast heating rate as well as by foreign matter in the solution, and this is particularly true at the higher temperatures and greater MEA concentrations. If foaming is detected, slightly reduce the burner fuel with the manual valve. This will reduce only the heating rate, not the final temperature as controlled by the modulating valve.

If the liquid temperature in the kettle is not increasing even though the fuel valve is fully open, take a sample of the solution in a beaker and let it stand for about 15 minutes until all of the air bubbles have disappeared. If the sample is clear, it indicates the fire tube surface in the kettle is clean. If the sample is white and cloudy with sediment, it is a good indication that the fire tubes need cleaning. A sample even though clear but not coming up to temperature may indicate internal carbon buildup as a result of incomplete combustion. Apply some Red Devil powder and adjust the air mixing valve. Internal fire-box carbon buildup may require removing the tube and burning the carbon deposit off by heating the opposite side of the tube walls. The need for cleaning a reclaimer system will vary from location to location, and only experience can determine the length of time between. Certainly, where historically the solution is highly loaded with contaminants, more frequent servicing will be required.

Many of the problems associated with an amine plant are immediately evident as they occur. Not so with corrosion, only microscopically evident in its initial advent. Corrosion probably incurs the greatest amine plant operating expense over a period of time. Fig. 5–11 is a schematic showing the most likely places for corrosion to occur as well as foaming, which also leads to corrosion if allowed to continue uninhibited. Most internal corrosion in an amine system occurs as a result of suspended solids deposited in areas of relatively low solution velocities. A type of corrosion resulting from this is galvanic cell corrosion, which leaves the metal surface pitted. Another type of corrosion occurs at points of high solution velocity. This is the result of a combination of corrosion-erosion. MEA in itself can be considered as a mild corrosion inhibitor because amine is an excellent film former; however, the film is delicate and will be destroyed by high velocities.

Stress corrosion refers to the susceptibility of materials to corrosive attack when under a state of stress. In an MEA desulfurization system, stress corrosion is related to residual or stresses that remain in the metal due to localized heating such as occurs in welds.

Some of the causes of corrosion have been noted. Following are some preventive measures, some which can be implemented through work location maintenance and operating procedure and others that may require further research and engineering evaluation. In these latter instances, "the pain, if endurable, may be preferable to the cure," the capital costs being excessive.

1. Close surveillance of acid-gas loading. This simply means *not* to exceed design solution strength, which in most instances is 20%, and to experiment by reducing the solution strength. Even a 1% reduction will contribute a great deal. It is not, however, suggested that it be reduced ever below 16% or so for a plant design of 20%. The reason for this precaution is that the reservoir could suddenly begin producing a greater amount of acid gas than when the lower solution concentration appeared to be working okay. Obviously, when reduced solution strength is used, the circulation rate may need to be increased. The danger in this is in upsetting the mechanical design criteria built into the plant.

2. Be sure the reboiler is operating at the lowest practical temperature, 230–240°F.
3. Use tougher materials such as stainless steel, where possible, when making repairs. Check this out with corrosion or metallurgical personnel before doing so because the application of dissimilar metals may provide a greater problem than the one being repaired.
4. Use every precaution to avoid any exposure of the solution to atmosphere to prevent oxidization. Be sure of adequate gas-blanket coverage for solution or amine in storage. Keep drums tightly sealed.
5. Good filtration.
6. Proper reclaimer operation.
7. Where possible, the rich solution should pass through the tube of heat exchangers and the lean in the shell. This will help to prevent corrosion-erosion of the fin tube, which is on the shell side of the tube.
8. Stress relieve all welds.
9. All solution that has been recovered from leaks, etc., or otherwise exposed to the atmosphere should be returned to the system through the reclaimer.
10. Heat exchange velocities should be held to a minimum. Some experts say 3 ft/sec, realizing that this is a product of plant design. However, it could also be a product of plant operation where the operator is attempting to adjust circulation rates to meet a lesser operating concentration of solution.

    Note: Velocity of flowing liquids in feet per second where gpm is known can be calculated:

$$V_{fps} = 0.4085 \ (gpm) \ /D^2$$

where:

$$
\begin{aligned}
V_{fps} &= \text{velocity, ft/sec} \\
0.4085 &= \text{a constant} \\
gpm &= \text{gal/min} \\
D &= \text{diameter of pipe, in.}
\end{aligned}
$$

Or using the previous example of 58 gpm and a 2-in. pipe:

$$V_{fps} = 0.4085 \ (58) \ /2^2$$
$$= 23.69/4$$
$$= 5.9, \text{ or approximately 6 ft/sec}$$

11. Use corrosion inhibitors. A primary requirement is that the inhibitor be water-soluble rather than the conventional hydrocarbon-soluble types. One such inhibitor is the Ucar process, marketed on a royalty basis by Union Carbide. However, under any condition the inhibitor must be totally compatible with other chemical compounds of the system: well inhibitors, impurities, etc. Using a particular inhibitor should be per recommendation of some authority capable of professionally analyzing the system and its requirements.

12. Avoid hot spots in fire-tube sections caused by the flame not being properly centered or, again, by foaming where the foam has replaced the liquid and consequently the foam instead of the liquid is in intimate contact with the outer fire-tube wall.

13. Sedimentary deposits tend to build up in areas of low velocity in the system, particularly in the reboiler, and are conducive to galvanic corrosion. Consideration should be given to tapping such locations and, if necessary, even installing a small centrifugal pump for recirculation.

Up to this point much has been said about foaming and its causes. Summarizing, the best cure for foaming problems is the proper care of the amine solution, inclusive of good practices and results in the following areas.

1. Inlet gas cleaning.
2. Filtration.
3. Observing contactor capacities.

4. Systematic observance of proper practices when placing in service the system or any part of the system.
5. Reclaimer operation.
6. Solution temperature control.
7. Use of foam inhibitor, making sure that the inhibitor used is compatible with the system. Like a corrosion inhibitor, recommendations for a foam inhibitor should be from someone qualified. Do not use more than recommended; too much can cause foaming. On-site tests should be made using a sample of solution and a sample of the inhibitor by observing its action in a glass beaker. Inject inhibitor exactly as the supplier recommends.
8. Do not use oils or greases (plug valve greases) that have a soap or detergent base.

Installing activated charcoal solution treating facilities is quite beneficial in most cases of severe foaming. Finally and for the operator to be aware of is that almost everything that causes corrosion will also cause foaming.

Many temperature-related conditions have been discussed here. Actual temperature conditions can only be specific where the reboiler and still column are concerned. Other temperature conditions are less specific, as they are affected to some extent by extraneous condition: weather, inlet gas temperature, heat exchanger efficiency, fan adjustment, etc. Typical temperatures for an MEA desulfurizer are:

Reboiler	230°F
Solution into contactor, lean	90°F
Solution into still, rich	185°F
Reflux into cooler	212°F
Reflux from cooler	130°F
Lean MEA to exchanger	140°F
Lean MEA from exchanger	90°F

One of the most demanding temperature requirements is that the contactor should be at a lesser temperature than the lean amine into the absorber.

Temperature is also critical to the storage of MEA, in that it freezes at 50.9°F and is completely soluble in water at 68°F. Obviously, MEA must be stored at a temperature around 55°F to be ready for use and to be in liquid rather than solid form. Freezing point of MEA solutions are:

% wt MEA in water	Freezing point
20%	30°F
30%	0°F
40%	−10°F

Solution stored in tanks at controlled temperatures should have a sweet gas blanket maintained to avoid oxidation.

Control of pH is exactly the same as required for TEG systems and should be coordinated through technically qualified personnel. Basically, if the solution is overloaded with acid gases (CO_2/H_2S), the amine tends to become neutral and swing the pH toward the corrosive acid side on the lower side of 7. Normally, the amine will be an off-white color, but as it tends to become acid it is a dirty gray or black in appearance because of iron oxide and iron sulfide contamination, indicating the occurrence of corrosion.

The final and probably the most important phase of MEA desulfurization plant operation in a storage field should occur between reservoir withdrawal periods—annual maintenance. During the operational period of the plant, a daily log should be kept, detailing trouble areas incurred—particularly leaks—during the operating season and the frequency at which each occurred. This will provide a starting place and help to establish maintenance priority. Also during the operating season and by consulting the aforementioned notes, it will become apparent that certain gaskets and spare parts should be ordered in anticipation of the maintenance required. Such a log will provide necessary data for construction budget requests. A maintenance and repair schedule should be prepared, taking into consideration the work to be accomplished, the number of personnel and working hours available, and alternate procedures in the event some planned work is interrupted.

The very first action to be taken is to continue circulating the solution with no gas flow through the contactor. (If it is possible to circulate sweet gas through the system during this period, it would be highly desirable even if only a small volume. This would provide some agitating action on the trays and help clean up.) Close either the inlet or outlet gas valve but maintain a static pressure with sweet gas. Operating temperatures should be reduced yet kept high enough that the solution does not become viscous. Continue to circulate, changing filters as often as indicated necessary, until the solution begins to clean up and the filters appear clean. After this has been done, the system should be evacuated of amine solution by storing in tanks under a sweet gas blanket. The remaining suspended sediment will tend to gravity out of the solution to the bottom of the storage tank where it can later be removed.

Next, the system should be filled with water and the circulation process repeated. During the water wash, solvents may be used but only those recommended by technical personnel. Many of the commercially available compounds leave a film that later produces foaming in routine operation, while other solvents may react with the amine or impurities and generate harmful or toxic vapors. Be sure that responsible personnel are aware of the concern. Upon completing the water wash, the system should be completely evacuated and previously noted trouble areas should be dismantled for inspection and repair.

The plant operating personnel and the experiences and troubles incurred during the past operating season provide the most valuable source of determining how extensive the dismantling inspection should be. There are, however, certain jobs that should be done at least annually. These include but are not limited to pulling the fire tubes in both the reboiler and reclaimer; pulling several fin-tube heat exchanger sections; complete inspection of all pumps; complete inspection of fuel and fire-box controls; complete inspection of all control valves and activator assemblies on both liquid and gas;

complete inspection of all safety devices, including alarms as well as shutdowns. Contactors should be inspected to the extent practical. Any sludge deposits should be removed and the metal closely inspected for corrosion. All vents should be inspected for corrosion.

One word of caution at this point: Do not attempt to reuse gaskets for the sake of economy. The probable solution losses incurred by reusing gaskets will normally pay for the new ones.

Any vessel that must be entered must be thoroughly purged and kept free of chemical vapors during the work period. All sources of gas must be totally isolated from the vessel. All vessels should be steam cleaned and purged before entering. Safety equipment and vapor-detecting devices should be available. Consideration should be given to laying down contactors and still columns when internal maintenance is required. Safety is all-inclusive of industry policy, with special consideration for the type of work to be performed. MEA, like TEG, is not in the hazardous materials classification; nevertheless, it is a chemical compound requiring proper judgment in handling.

Trial runs on desulfurization plants should be made well ahead of anticipated onstream time to determine if further repairs are needed.

In conclusion, the degree of trouble-free operation during any withdrawal period will be about proportional to the off-season maintenance performed and the establishment and pursuit of good operating practices while the plant is in operation.

PART VI TYPICAL OPERATING FLOW DIAGRAMS

The following flow diagrams are typical of dehydration and desulfurization facilities utilizing TEG and MEA solutions, the former being primary and the latter in series. They illustrate the physical relationship of each to the other and typify operating conditions.

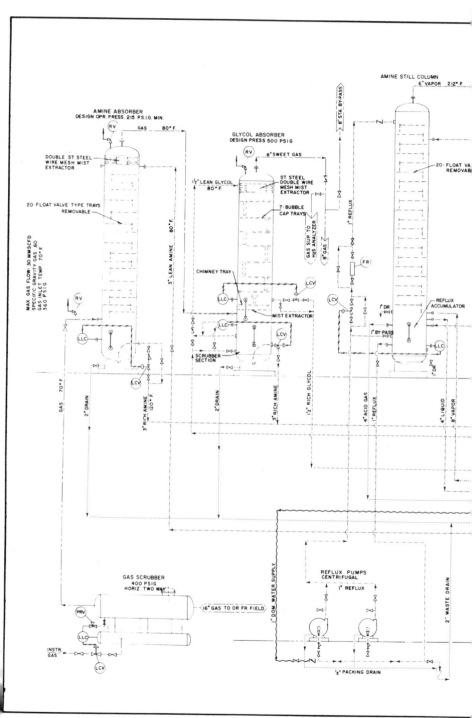

6–1   Typical plant #1, 30 MMcf/day @ 360 psig

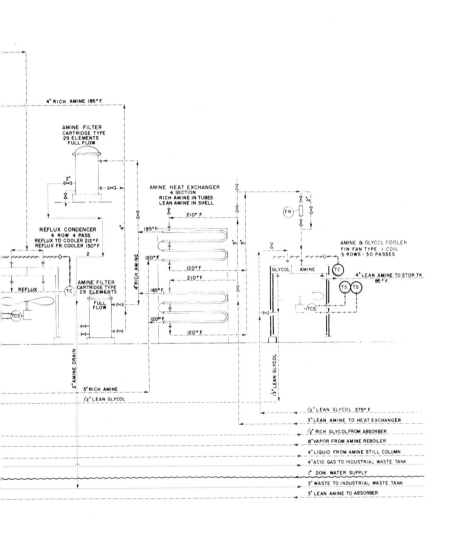

4" RICH AMINE 185° F.

AMINE FILTER
CARTRIDGE TYPE
29 ELEMENTS
FULL FLOW

2"

AMINE HEAT EXCHANGER
4 SECTION
RICH AMINE IN TUBES
LEAN AMINE IN SHELL

210° F.

185°F.

FR

REFLUX CONDENCER
4 ROW 4 PASS
REFLUX TO COOLER 212° F.
REFLUX FR. COOLER 130° F.

120°F.

AMINE & GLYCOL COOLER
FIN FAN TYPE I COIL
5 ROWS - 30 PASSES

120° F.

210° F.

GLYCOL   AMINE

TC

AMINE FILTER
CARTRIDGE TYPE
29 ELEMENTS
FULL FLOW

185°F.

4" LEAN AMINE TO STOR TK
85° F.

REFLUX

TC

TS   TS

120° F.

TCS

TCS

120° F.

2" AMINE DRAIN

3" RICH AMINE

1½" LEAN GLYCOL

1½" LEAN GLYCOL

1½" LEAN GLYCOL 375° F.

3" LEAN AMINE TO HEAT EXCHANGER

1½" RICH GLYCOL FROM ABSORBER

8" VAPOR FROM AMINE REBOILER

4" LIQUID FROM AMINE STILL COLUMN

4" ACID GAS TO INDUSTRIAL WASTE TANK

1" DOM. WATER SUPPLY

3" WASTE TO INDUSTRIAL WASTE TANK

3" LEAN AMINE TO ABSORBER

TYPICAL PLANT NO. 1

FLOW DIAGRAM

SHEET 1 OF 2

30MMcfd @ 360 psig.

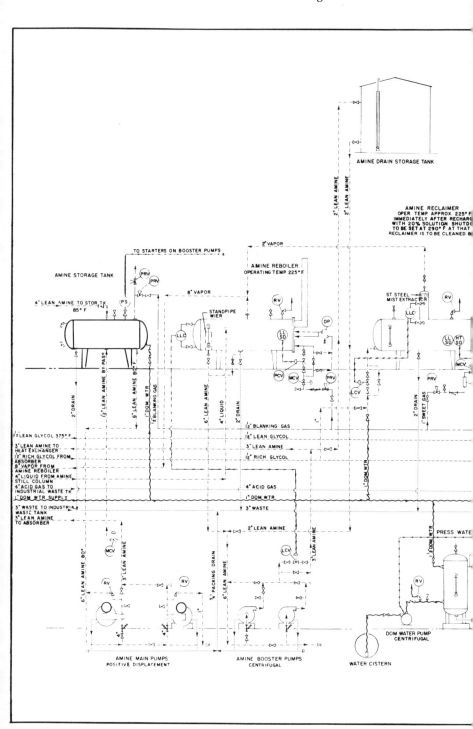

6–2 Typical plant #1, *continued*

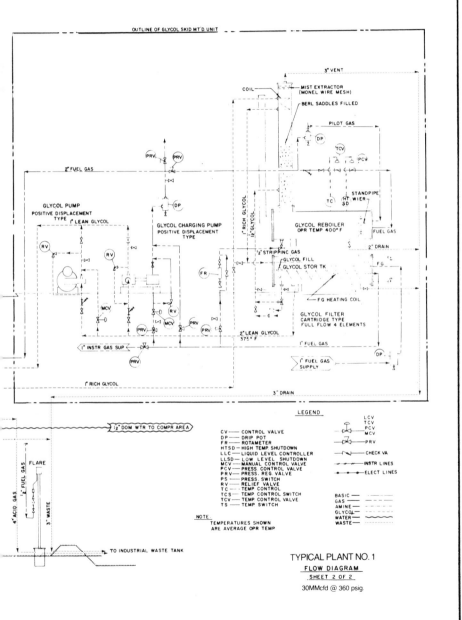

LEGEND

CV — CONTROL VALVE
DP — DRIP POT
FR — ROTAMETER
HTSD — HIGH TEMP. SHUTDOWN
LLC — LIQUID LEVEL CONTROLLER
LLSD — LOW LEVEL SHUTDOWN
MCV — MANUAL CONTROL VALVE
PCV — PRESS. CONTROL VALVE
PRV — PRESS. REG VALVE
PS — PRESS. SWITCH
RV — RELIEF VALVE
TC — TEMP CONTROL
TCS — TEMP CONTROL SWITCH
TCV — TEMP CONTROL VALVE
TS — TEMP SWITCH

LCV
TCV
PCV
MCV

——— PRV

——— CHECK VA

—— INSTR LINES
—•—•— ELECT LINES

BASIC ———
GAS ———
AMINE ———
GLYCOL ———
WATER ———
WASTE ———

NOTE.
TEMPERATURES SHOWN
ARE AVERAGE OPR TEMP

TYPICAL PLANT NO. 1
FLOW DIAGRAM
SHEET 2 OF 2
30MMcfd @ 360 psig.

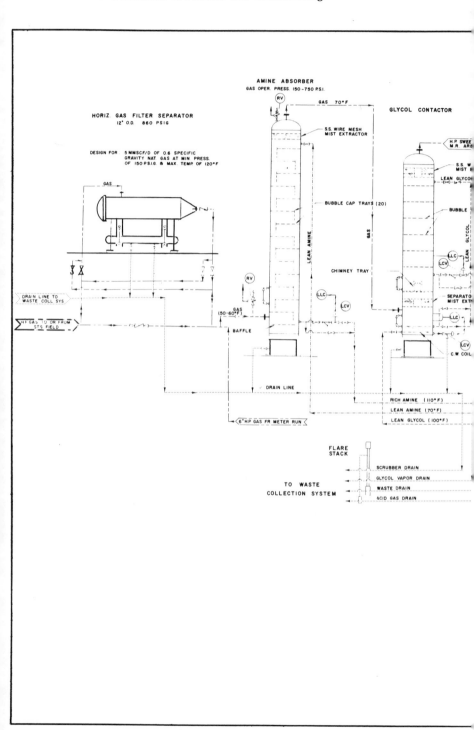

6–3 Typical plant #2, 5 MMcf/day @ 150 psig/min

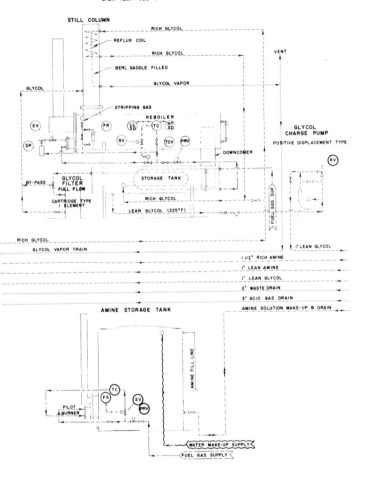

GLYCOL REGENERATION UNIT
OPER. TEMP. 400° F.

STILL COLUMN

RICH GLYCOL

REFLUX COIL

RICH GLYCOL

VENT

BERL SADDLE FILLED

GLYCOL VAPOR

GLYCOL

STRIPPING GAS

REBOILER

GLYCOL
CHARGE PUMP
POSITIVE DISPLACEMENT TYPE

DOWNCOMER

GLYCOL
FILTER
FULL FLOW

BY-PASS

STORAGE TANK

CARTRIDGE TYPE
ELEMENT

RICH GLYCOL

LEAN GLYCOL (225°F)

RICH GLYCOL

GLYCOL VAPOR DRAIN

1" LEAN GLYCOL

1·1/2" RICH AMINE

1" LEAN AMINE

1" LEAN GLYCOL

2" WASTE DRAIN

3" ACID GAS DRAIN

AMINE SOLUTION MAKE-UP & DRAIN

AMINE STORAGE TANK

AMINE FILL LINE

PILOT
BURNER

WATER MAKE-UP SUPPLY

FUEL GAS SUPPLY

TYPICAL PLANT NO. 2

FLOW   DIAGRAM

SHEET   1 OF 2

5MMcfd @ 150 psig. min.

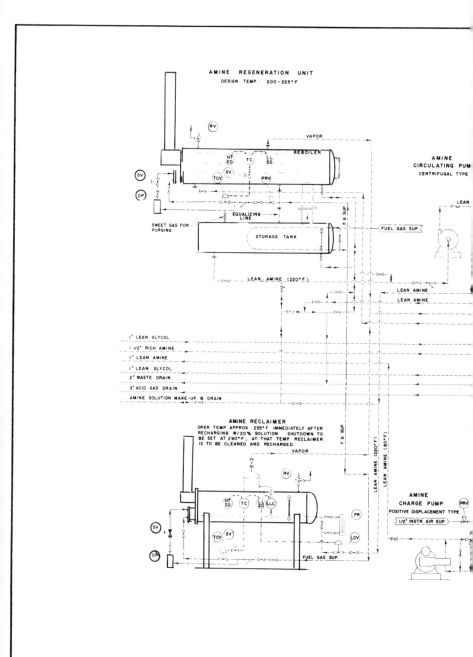

6–4 Typical plant #2, *continued*

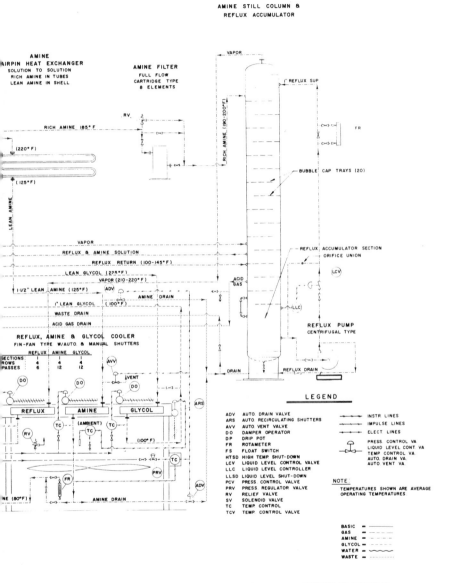

AMINE STILL COLUMN &
REFLUX ACCUMULATOR

AMINE
HAIRPIN HEAT EXCHANGER
SOLUTION TO SOLUTION
RICH AMINE IN TUBES
LEAN AMINE IN SHELL

AMINE FILTER
FULL FLOW
CARTRIDGE TYPE
8 ELEMENTS

VAPOR

1" REFLUX SUP

FR

RICH AMINE 185°F

RV

BUBBLE CAP TRAYS (20)

(220°F)

(125°F)

VAPOR
REFLUX & AMINE SOLUTION
REFLUX RETURN (100-145°F)
LEAN GLYCOL (225°F)
VAPOR (210-220°F)
1-1/2" LEAN AMINE (125°F)   ADV   AMINE DRAIN
1" LEAN GLYCOL (100°F)
WASTE DRAIN
ACID GAS DRAIN

REFLUX ACCUMULATOR SECTION
ORIFICE UNION

LCV

ACID
GAS

LLC

REFLUX PUMP
CENTRIFUGAL TYPE

REFLUX, AMINE & GLYCOL COOLER
FIN-FAN TYPE W/AUTO & MANUAL SHUTTERS

SECTIONS	REFLUX	AMINE	GLYCOL
ROWS	1	1	1
PASSES	4	4	4
	6	12	12

AVV

VENT

DO   DO   DO

DRAIN   REFLUX DRAIN

ARS

REFLUX   AMINE   GLYCOL

TC (AMBIENT) TC

RV   TC

(100°F)

TC

PRV

FR

ADV

AMINE (80°F)   AMINE DRAIN

LEGEND

ADV	AUTO. DRAIN VALVE
ARS	AUTO. RECIRCULATING SHUTTERS
AVV	AUTO. VENT VALVE
DO	DAMPER OPERATOR
DP	DRIP POT
FR	ROTAMETER
FS	FLOAT SWITCH
HTSD	HIGH TEMP SHUT-DOWN
LCV	LIQUID LEVEL CONTROL VALVE
LLC	LIQUID LEVEL CONTROLLER
LLSD	LIQUID LEVEL SHUT-DOWN
PCV	PRESS. CONTROL VALVE
PRV	PRESS. REGULATOR VALVE
RV	RELIEF VALVE
SV	SOLENOID VALVE
TC	TEMP. CONTROL
TCV	TEMP. CONTROL VALVE

INSTR LINES
IMPULSE LINES
ELECT LINES
PRESS CONTROL VA
LIQUID LEVEL CONT VA
TEMP CONTROL VA.
AUTO. DRAIN VA.
AUTO. VENT VA.

NOTE
TEMPERATURES SHOWN ARE AVERAGE
OPERATING TEMPERATURES

BASIC =
GAS =
AMINE =
GLYCOL =
WATER =
WASTE =

TYPICAL PLANT NO. 2
FLOW DIAGRAM
SHEET 2 OF 2

5MMcfd @ 150 psig. min.

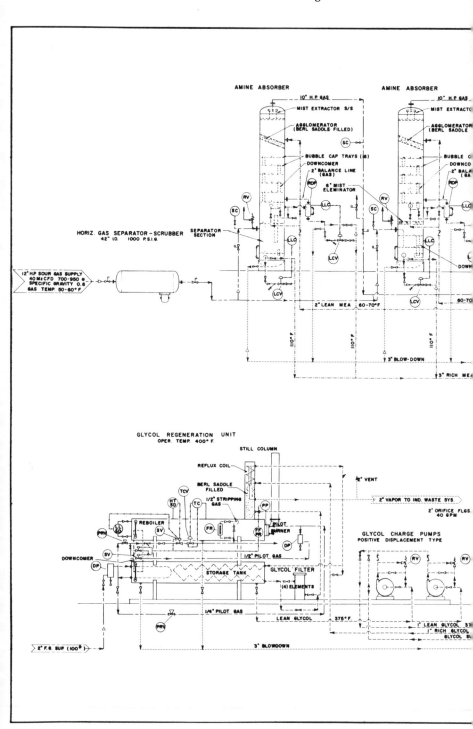

6–5 Typical plant #3, 40 MMcf/day @ 700–950 psig

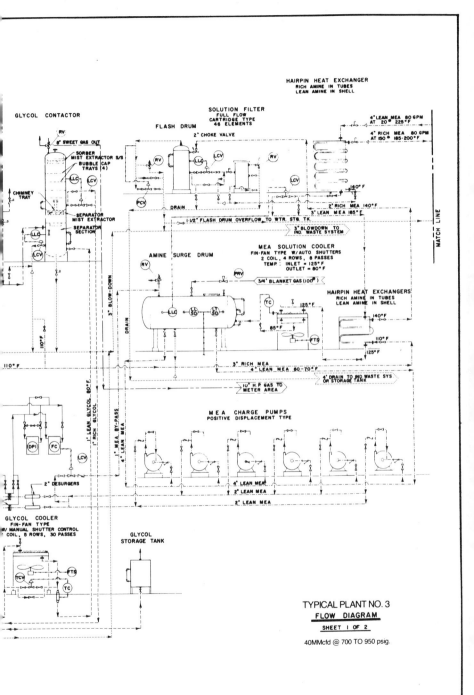

HAIRPIN HEAT EXCHANGER
RICH AMINE IN TUBES
LEAN AMINE IN SHELL

GLYCOL CONTACTOR

SOLUTION FILTER
FULL FLOW
CARTRIDGE TYPE
46 ELEMENTS

FLASH DRUM

4" LEAN MEA 80 GPM
AT 20° 225°F

4" RICH MEA 80 GPM
AT 150° 185-200°F

2" CHOKE VALVE

RV

8" SWEET GAS OUT

SORBER
MIST EXTRACTOR S/S
BUBBLE CAP
TRAYS (4)

CHIMNEY
TRAY

SEPARATOR
MIST EXTRACTOR

SEPARATOR
SECTION

140°F

2" RICH MEA 140°F

3" LEAN MEA 185°F

1 1/2" FLASH DRUM OVERFLOW TO WTR STG TK

3" BLOWDOWN TO
IND. WASTE SYSTEM

AMINE SURGE DRUM

MEA SOLUTION COOLER
FIN-FAN TYPE W/AUTO SHUTTERS
2 COIL, 4 ROWS, 8 PASSES
TEMP : INLET = 125°F
OUTLET = 80°F

3/4" BLANKET GAS (100°)

HAIRPIN HEAT EXCHANGERS
RICH AMINE IN TUBES
LEAN AMINE IN SHELL

125°F

140°F

85°F

110°F

125°F

110°F

3" RICH MEA

4" LEAN MEA 60-70°F

4" DRAIN TO IND WASTE SYS
OR STORAGE TANK

110°F

2" LEAN GLYCOL 80°F
1" RICH GLYCOL

10" H.P GAS TO
METER AREA

MEA CHARGE PUMPS
POSITIVE DISPLACEMENT TYPE

1" MEA BY-PASS
4" LEAN MEA

2" DESURGERS

4" LEAN MEA

2" LEAN MEA

2" LEAN MEA

GLYCOL COOLER
FIN-FAN TYPE
W/ MANUAL SHUTTER CONTROL
COIL, 5 ROWS, 30 PASSES

GLYCOL
STORAGE TANK

MATCH LINE

TYPICAL PLANT NO. 3
FLOW DIAGRAM
SHEET 1 OF 2

40MMcfd @ 700 TO 950 psig.

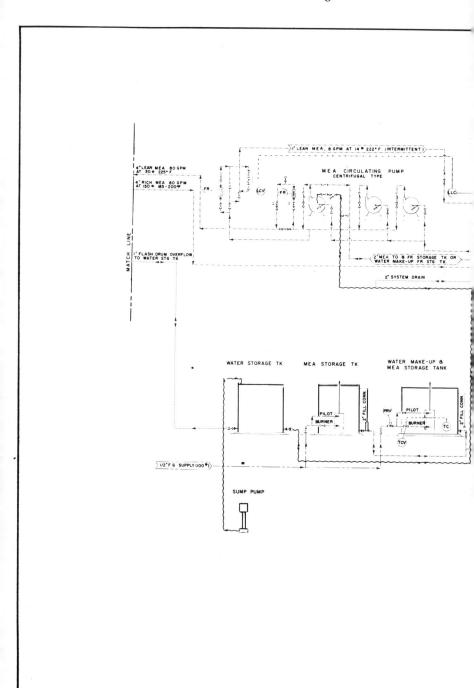

6–6 Typical plant #3, *continued*

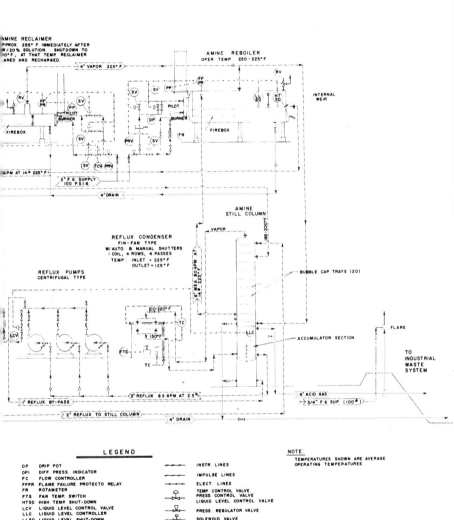

AMINE RECLAIMER
PPROX 255° F. IMMEDIATELY AFTER
W/20% SOLUTION. SHUTDOWN TO
0°F., AT THAT TEMP. RECLAIMER
ANED AND RECHARGED.

4" VAPOR 225° F

AMINE REBOILER
OPER TEMP 220-225°F

INTERNAL
WEIR

FIREBOX

PILOT
BURNER

PILOT
BURNER

FIREBOX

GPM AT 14° 225° F

2" F.G. SUPPLY
100 P.S.I.G.

4" DRAIN

AMINE
STILL COLUMN

REFLUX CONDENSER
FIN-FAN TYPE
W/ AUTO. & MANUAL SHUTTERS
I COIL, 4 ROWS, 4 PASSES
TEMP : INLET = 225° F
        OUTLET = 125° F

VAPOR

185-200°F

REFLUX PUMPS
CENTRIFUGAL TYPE

BUBBLE CAP TRAYS (20)

210-220° F

130°F

FLARE

TO
INDUSTRIAL
WASTE
SYSTEM

ACCUMULATOR SECTION

3" REFLUX 83 GPM AT 23"

1" REFLUX BY-PASS

6" ACID GAS

3/4" F.G. SUP (100 #)

2" REFLUX TO STILL COLUMN

4" DRAIN

LEGEND

DP	DRIP POT
DPI	DIFF. PRESS. INDICATOR
FC	FLOW CONTROLLER
FFPR	FLAME FAILURE PROTECTO RELAY
FR	ROTAMETER
FTS	FAN TEMP. SWITCH
HTSD	HIGH TEMP. SHUT-DOWN
LCV	LIQUID LEVEL CONTROL VALVE
LLC	LIQUID LEVEL CONTROLLER
LLSD	LIQUID LEVEL SHUT-DOWN
LTSD	LOW TEMP. SHUT-DOWN
PCV	PRESS. CONTROL VALVE
PRV	PRESS. REGULATOR VALVE
RDP	RUN DOWN POT
RV	RELIEF VALVE
SC	SENSING CONN.
SV	SOLENOID VALVE
TC	TEMP. CONTROL
TCV	TEMP. CONTROL VALVE

INSTR. LINES
IMPULSE LINES
ELECT. LINES
TEMP. CONTROL VALVE
PRESS. CONTROL VALVE
LIQUID LEVEL CONTROL VALVE
PRESS. REGULATOR VALVE
SOLENOID VALVE

BASIC =
GAS =
AMINE =
GLYCOL =
WATER =
WASTE =

NOTE:
TEMPERATURES SHOWN ARE AVERAGE
OPERATING TEMPERATURES

TYPICAL PLANT NO. 3
FLOW DIAGRAM
SHEET 2 OF 2
40MMcfd @ 700 TO 950 psig.

BIBLIOGRAPHY

Centrifix, Division of Burgess Industries, Bulletin 174-C, 1953.

Connors, J. S. (Phillips Petroleum Co.), Selected Proceedings from Gas Conditioning Conferences, III, University of Oklahoma.

Dingman, J. C., and T. F. Moore (Jefferson Chemical Co. Inc.), Selected Proceedings from Gas Conditioning Conferences, III, University of Oklahoma.

Gas Engineers' Handbook, Industrial Press, 1977.

Holm, E. J. (Northern Natural Gas Co.), Selected Proceedings from Gas Conditioning Conferences, III, University of Oklahoma.

Jones, V. Wayne, and Charles R. Perry (Perry Gas Processors), Selected Proceedings from Gas Conditioning Conferences, III, University of Oklahoma.

Peco Robinson, Gas Filters & Filter Separators, HA/9/75/5M.

Peerless Filter Separators, Bulletins 6-30 and 6-193.

Polderman, L. D. (Carbide Chemicals Co.), Selected Proceedings from Gas Conditioning Conferences, III, University of Oklahoma.

Redus, Frank R. (Colorado Interstate Gas Co.), Selected Proceedings from Gas Conditioning Conferences, III, University of Oklahoma.

Shell, Allen D. (Mobil Oil Corp.), *Oil & Gas Journal*, 26 February 1968.

Sivalls, C. Richard (Sivalls Tanks Inc.), Selected Proceedings from Gas Conditioning Conferences, III, University of Oklahoma.

U.S. Bureau of Mines, Monograph 8.

Wonder, D. K., et al. (Union Carbide Chemical Co.), Selected Proceedings from Gas Conditioning Conferences, III, University of Oklahoma.

SOURCES OF ILLUSTRATIONS

2–1   University of Oklahoma, Selected Proceedings from Gas Conditioning Conferences, Vol. 3, p. 3–29; paper by E.J. Holm. Northern Natural Gas Company.

2–2   Peerless Filter Separators, Bulletin 6–30, 1963.

2–3   King Tool, Drawing GSC–15758G.

2–4   Centrifix, Division of Burgess Industries, Bulletin 174–C, 1953.

2–5   Perrless In-Line, Bulletin 7–17S Model VGF.

2–6   Peco Robinson, Gas Filters and Filter Separators, HA/9/75/5M.

2–7   Perrless, Bulletin 6–193, Fig. 2.

4–1   U.S. Bureau of Mines, Monograph 8.

4–2   U.S. Bureau of Mines, Monograph 8.

4–3   Welker Engineering Company.

4–4   University of Oklahoma. Selected Proceedings from Gas Conditioning Conferences Vol. 3, p. 7–17; paper by L.D. Polderman Carbide Chemicals Company.

4–5   University of Oklahoma Selected Proceedings from Gas Conditioning Conferences Vol. 3, p. 9–11; paper by Frank R. Redus, Colorado Interstate Gas Company.

4–6   Black, Sivalls, Bryson service manual.

4–7   Black, Sivalls, Bryson service manual.

4–8   University of Oklahoma, Selected Proceedings from Gas Conditioning Conferences Vol. 3, paper by C. Richard Sivalls, Sivalls Tanks Inc.

4–10   Black, Sivalls, Bryson service manual.

4–12   University of Oklahoma, Gas Processors Equipment Short Course, sponsored by Carbide and Carbon Chemical Company.

4–13 University of Oklahoma, Gas Processors Equipment Short Course, sponsored by Magnolia Petroleum Company.

4–14 University of Oklahoma, Gas Processors Equipment short course.

5–2 University of Oklahoma, Selected Proceedings from Gas Conditioning Conferences, Vol. 3, p. 12–9; paper by J.S. Connors, Phillips Petroleum Company.

5–3 University of Oklahoma, Selected Proceedings from Gas Conditioning Conferences, Vol. 3, p. 18–9; paper by J.C. Dingman, T.F. Moore, Jefferson Chemical Company Inc.

5–4 University of Oklahoma, Selected Proceedings from Gas Conditioning Conferences, Vol. 3, p. 19–12; paper by J.C. Dingman, T.F. Moore, Jefferson Chemical Company Inc.

5–5 University of Oklahoma, Selected Proceedings from Gas Conditioning Conferences, Vol. 3, p. 23–8; paper by V. Wayne Jones, Charles R. Perry, Perry Gas Processors.

5–6 *Gas Engineer's Handbook*, Industrial Press, 1977.

5–7 University of Oklahoma, Selected Proceedings from Gas Conditioning Conferences. Vol. 3, p. 15–1; paper by D.K. Wonder et al., Union Carbide Chemical Company.

5–8 *Oil & Gas Journal*, 26 February 1968, Allen D. Shell, Mobil Oil Corporation.

5–9 University of Oklahoma, Selected Proceedings from Gas Conditioning Conferences, Vol. 3, p. 12–15; paper by J.S. Connors, Phillips Petroleum Company.

5–10 University of Oklahoma, Selected Proceedings from Gas Conditioning Conferences, Vol. 3, p. 14–17; paper by D.K. Wonder et al., Union Carbide Chemical Company.

5–11 *Oil & Gas Journal*.

INDEX